Praise for *The Burning Answer*

'Barnham does have something new to say: he cuts through the current morass of fossil-fuel and nuclear lobbyists' negative propaganda with a clear and original vision for solar power . . . It is a bold vision, a necessary one, and the world needs to be fired up about it. Keith Barnham is fanning a necessary flame'

Peter Forbes, *Guardian*

'The overwhelming impression I take away from *The Burning Answer* is one of a slowly building but completely unstoppable momentum behind this solar revolution . . . It's one of the most exciting and genuinely hopeful books I've read in a long time'

Jonathon Porritt, *Ecologist*

'A scorching diatribe . . . the energy fizzes off the page, and any doubters of the merits of solar power, be they scientists, commentators or policy wonks, are given a swift left-hook'

Nicola Davis, *Observer*

'An excellent resource in explaining the science, historical development and future potential of solar technologies . . . Barnham offers steps we can take individually and collectively as "solar revolutionaries"'

Helen Moore, *Permaculture*

'This is a very important book . . . Here we have an actual physicist, the kind who worked on honking big particle accelerators, breaking down the underlying science . . . It is with the assurance of a craftsman, as well as a theoretician, that he undertakes to demonstrate that renewable energy is indeed capable of doing what needs to be done when it comes to the real world of toast, watching Netflix, and getting you to work'

Bill McKibben, author of *The End of Nature*

Keith Barnham is Emeritus Professor of Physics at Imperial College London. He started his research career in experimental particle physics working in research laboratories at CERN in Geneva and the University of California, Berkeley. Mid-career he switched to researching solar energy and invented a solar cell with three times the efficiency of today's roof-top panels. He lives with his wife, the poet Claire Crowther, in Somerset.

THE
BURNING
ANSWER

A User's Guide to the Solar Revolution

KEITH BARNHAM

W&N
WEIDENFELD & NICOLSON

A W&N PAPERBACK

First published in Great Britain in 2014
by Weidenfeld & Nicolson
This paperback edition published in 2015
by Weidenfeld & Nicolson,
an imprint of the Orion Publishing Group Ltd,
Carmelite House, 50 Victoria Embankment,
London, EC4Y 0DZ

An Hachette UK company

1 3 5 7 9 10 8 6 4 2

A CIP catalogue record for this book
is available from the British Library.

ISBN 978-1-7802-2533-3

Typeset by GroupFMG within BookCloud
Printed and bound in Great Britain by
CPI Group (UK) Ltd, Croydon, CR0 4YY

The Orion Publishing Group's policy is to use papers that
are natural, renewable and recyclable products and
made from wood grown in sustainable forests. The logging
and manufacturing processes are expected to conform to
the environmental regulations of the country of origin.

www.orionbooks.co.uk

Judith Barnham
17-5-42 to 27-11-94
"Her life of light and love"

CONTENTS

Preface

When the first edition of *The Burning Answer* was published last year, I wanted to leave the reader with a vision of a solar future in which the sun supplies all our energy needs. Gone would be the threats posed by global warming, nuclear disaster and conflict in oil-rich countries, all three of which are linked to our thirst for energy.

A year later, the solar revolution has moved on apace. There is already a need to update the information in Part II: 'The Here and Now of the Solar Revolution'. The first exciting new development emerged when I looked up the latest figures on renewable electricity installations published by the UK Department of Energy and Climate Change (DECC). I decided to plot a graph of the rate at which offshore wind power was expanding, in the same way that I have tracked the growth of photovoltaic (PV) installations for more than ten years. Incredibly, offshore wind power installations around the UK have been expanding at a rate not far short of that achieved by PV in Germany in the halcyon years of their feed-in tariff. This rapid expansion is typical of small semiconductor devices, like mobile phones, that can be prototyped and mass produced quickly. But offshore wind turbines are massive structures of novel design, installed in a challenging environment. Such a rapid expansion of wind power is an extremely impressive achievement.

If this rate of expansion continues, the UK could achieve the level of offshore wind contribution required for an all-renewable electricity supply soon after 2020. Part II describes how PV installations could also achieve their level of contribution needed for an all-renewable UK electricity supply by a similar date.

The latest news from Germany is also positive. The fall in the peak wholesale price of electricity, described in Part II, has continued. There is now so much PV and wind power on the German grid that the average wholesale electricity price, not just the peak price, has declined in recent years.

These developments contrast dramatically with the lack of progress on new nuclear reactors. The two prototypes for the European Power Reactor (EPR) planned at Hinkley Point are still years away from operating. If offshore wind and PV maintain their current rates of expansion, they will each provide well over 20 times as much power as one EPR before the latter's earliest operation date of 2023.

In the past year, the European Commission has approved the UK government's subsidy scheme for nuclear reactors. This guarantees that the public will fund the difference between a fixed, high price for nuclear electricity and the wholesale price for electricity in a decade's time. The large amount of renewable electricity on the grid by 2023 will mean the wholesale price will be lower than now, so the subsidy will be higher.

The Commission's decision is likely to be appealed. The new UK government, EDF and the investors should use this period for reflection and follow the advice in this book. Good engineering and investment practice requires them to await the operation of one of the EPR prototypes before signing a contract.

The Burning Answer shows that renewable alternatives to Hinkley Point C are cheaper, have a lower carbon footprint and can be built much faster. Why the UK government is so obsessed with nuclear power is one of a number of political and scientific mysteries that the book sets out to solve.

One set-back for the solar revolution in 2014 was the UK coalition government's decision to remove the subsidy for solar farms. The government argued that solar farms are taking too large a share of the renewables subsidy. Why didn't the government simply transfer some of the much larger fossil fuel subsidies to PV? Solar farms will now have to compete with other renewable technologies, and with nuclear, for a finite subsidy pot from which expensive nuclear will take the lion's share. Such uncertainties in funding can be fatal for young solar companies.

There were other set-backs for the solar revolution in 2014. Both the Conservatives and UKIP voiced opposition to onshore wind power. A coalition government minister recorded his fiftieth rejection of onshore wind projects after they had been approved by local government.

For every depressing news story in the past year there has been at least one, less well-publicised, success story for the solar revolution. The decision to cut the solar farm subsidy coincided with DECC publishing results from public opinion surveys that show around 70 to 85 per cent of the UK support wind and solar power. Most significantly, when respondents were asked if they would be happy with 'a large scale renewable energy development' in their area, this support did not fall by much (59 per cent were still in favour). The support for wind and solar power was considerably higher than the support for nuclear (42 per cent) and more than double the support for fracking (29 per cent). No doubt the percentages favouring nuclear and fracking would have fallen even further, had respondents been asked if they would like nuclear reactors or fracking rigs in their neighbourhood.

There has also been excellent news on biomethane. Part II describes how the first anaerobic digestion (AD) plant in the UK, which produces biomethane from agricultural and food waste, was connected to the gas grid in 2012. Ciaran Burns of the Greengas certification scheme, which tracks the biomethane after injection into the grid, updated the situation for me. By the end of 2014, twenty-eight plants were in operation and these were capable of injecting enough biomethane to heat 100,000 UK homes. There should now be sufficient biomethane for companies that provide all-renewable electricity to offer renewable gas as well.

Representatives of both Ecotricity and Good Energy explained that they hope to be able to offer customers biomethane in 2016.

There are other exciting applications of biomethane from AD. Christopher Maltin, director of Biomethane Ltd, explained to me that it has benefits for gas-powered lorries, buses and cars, and Grant Ashton from the Biomethane Certification Scheme reported that demand for their certificates from transport companies has recently taken off.

The solar revolution is now expanding outside Europe. India has announced plans for 100 GW of PV by 2022. China is expected to install nearly 18 GW of PV in a single year in 2015. The charity, SolarAid, has already passed the milestone of 1.5 million solar lights sold in Africa. They are on target for their goal of eradicating oil use for lighting on the African continent by 2020. After writing an article for the *Ecologist*, which suggested how renewables could ensure a lasting ceasefire in Gaza, I heard good news from the Sunshine4Palestine charity. They have already installed PV at a hospital in Gaza, dramatically increasing the time electrical medical equipment can be used each day.

In the build-up to the December 2015 UN Climate Conference in Paris, NGOs and internet campaign groups have organised impressive public demonstrations in many countries calling on governments to agree to significant carbon cuts. *The Burning Answer* suggests that the most effective policy would be for governments to set an early date at which all new electricity generators will have greenhouse gas emissions below 50 gCO_2/kWh.

For me, one of the greatest pleasures of the past year has been meeting the workers at the grass roots (rather than the coalface) of the solar revolution. I have travelled around the UK, mainly by train, from Belfast in Northern Ireland to Dorchester in the south to talk to local environmental groups. The enthusiasm and determination of the solar revolutionaries I have met is impressive.

In Belfast, at the Northern Ireland Energy Forum, it was exciting to see solar overcoming old political divisions. The co-operation between Northern Ireland and Eire on improving grid connection will benefit both parts of Ireland. The north has superior wind power and the south has better PV resources. I also heard from Regan Smyth that the Friends of the Earth 'Run on Sun' project has already encouraged one-third of Northern Ireland schools to install solar panels.

Visiting the western mountains of Wales, I was greatly encouraged by the work on Zero Carbon Britain by the Centre for Alternative Technology. In deepest Somerset, appropriately in the part of the county affected by flooding in 2014, Eva Bishop of Tidal Lagoon Power talked about their planned installation in Swansea Bay. A tidal lagoon off Hinkley Point could generate more power than the two planned EPRs together, and operate earlier. Its power generation is completely predictable. This would complement the offshore wind and PV power proposed as alternatives to Hinkley Point C in Part III. Peter Lee, retired harbour master at Burnham-on-Sea near Hinkley Point, explained how a lagoon would also alleviate flooding on the Somerset levels.

All the groups I have met in the past year have strengthened my belief in the ultimate success of the solar revolution. I believe there is now grass roots activity in the UK comparable with that in the pioneering countries of Denmark, Germany and Sweden. I sense this momentum is also building in countries outside Europe.

I hope you will agree with the environmentalist Jonathon Porritt that 'the overwhelming impression I take away from *The Burning Answer* is one of a slowly building but completely unstoppable momentum behind this solar revolution'.

April, 2015

Introduction

This book contains a manifesto for a solar revolution. Not so long ago the sun provided all our energy needs. This book will explain how we can return to that situation. The revolution is already well underway in a few countries. We can all participate in it, and together we can persuade more governments, and even the fossil fuel and nuclear industries, to join us.

I will show you how the renewable technologies exploit the energy generated by our magnificently burning sun. They offer us the best chance to avoid the triad of threats posed by global warming, fossil fuel depletion and the devil's bargain of nuclear power. All three dangers arise because we produce most of our energy by burning unsustainable fuel. News stories keep reminding us of the seriousness of these threats: reports of shrinking ice caps, famine in East Africa, a hurricane in New York, thousands killed by a typhoon in the Philippines, conflict in countries on which we depend for oil and the proliferation of nuclear weapons to North Korea and possibly Iran. Environmental disasters have occurred in the Gulf of Mexico and downwind of the nuclear reactors in Fukushima. There is environmental damage from coal burning, fracking, shale-oil extraction, deforestation and drilling for fossil fuels in sensitive environments.

E = mc²: There is an alternative

Stephen Hawking, the famous cosmologist, made an important conjecture about one of these three existential threats when he speculated about alien civilisations elsewhere in the cosmos. He posed the question: if more advanced civilisations do exist somewhere in the cosmos, why have none of them yet colonised our earth?

Hawking's answer is that they have all discovered the equation $E = mc^2$, which is the key to both nuclear power and nuclear weapons, and blown themselves up.

Is our civilisation destined to come to a similar, catastrophic end? It does not have to; there is an alternative source of power. I am going to tell you about another equation $E = hf$, which is much less well known. It is the equation that started the quantum revolution, which in turn led to the semiconductor revolution that provided your laptop and mobile phone. This equation is also fundamental to the generation of electricity from sunlight. The technology of the solar revolution, which is based on $E = hf$, can save us from blowing ourselves up with $E = mc^2$.

Whether or not Hawking is right and civilisations exist that discovered $E = mc^2$ and blew themselves up, it surely is equally possible that there are civilisations in the cosmos that discovered and exploited $E = hf$ before they discovered $E = mc^2$. If these civilisations did realise that $E = hf$ is the key to sustainable, conflict-free, energy generation, they probably did not need to exploit $E = mc^2$ and so have not blown themselves up. Such civilisations have no cause to colonise earth. They have not overheated their planet by burning. Their oil and uranium resources are not depleting. They have no need to dump their nuclear waste on earth.

I hope to explain enough about the application and the physics of both equations to demonstrate an amazing coincidence about humankind. Our civilisation is on the cusp. We are possibly the only civilisation in the universe that did not discover one or other of these crucial equations first. In our civilisation, the equation $E = mc^2$ was discovered and the equation $E = hf$ was explained *in the same year by the same man – Albert Einstein.*

A century later, our civilisation could still go either way; destroying ourselves with $E = mc^2$ like all alien civilisations as Hawking speculates, or achieving sustainability with $E = hf$ like civilisations that may have survived.

I want to show you how the equation that can save our civilisation works. In fact, few popular science books have much to say about $E = hf$, though there are many that explain $E = mc^2$. Recently, Brian Cox and Jeff Forshaw have written an entire book about $E = mc^2$.

There is an example of the view among some scientists that $E = mc^2$ is more important than $E = hf$ in Bill Bryson's popular book *A Short History of Nearly Everything*. Like many people, I am a big fan of Bill Bryson's travel writing. I am also in awe of his achievement in making such a wide range of scientific issues so accessible. However, one problem for any non-specialist writer is that, when it comes to assessing priorities, scientific advisers often reflect differing views. In no way do I wish to criticise Bill Bryson; it is simply that my opinion on the relative importance of Einstein's achievements differs from that of some of Bryson's mentors.

Here is Bill Bryson on the three great papers Einstein published in 1905 (the first being about $E = hf$ and the third about relativity):

> The first won its author a Nobel Prize and explained the nature of light (and also helped to make television possible, among other things). The second provided proof that atoms do indeed exist – a fact that had, surprisingly, been in some dispute. The third merely changed the world.

In my view, Einstein's first paper, which explained $E = hf$, changed the world at least as much as relativity. He wrote a fourth paper in 1905 that shows how relativity leads to the equation $E = mc^2$. I will claim that his new explanation of the nature of sunlight was revolutionary. Furthermore, television got on very well for many years without the help of the quantum theory which resulted from $E = hf$. It was only when TV sets and TV cameras started getting smaller that Einstein's $E = hf$ became useful. Today's flat screens would have been impossible without this equation.

When I describe the 'other things' that developed from Einstein's first paper – the silicon chip, personal computer, digital camera, mobile phone, e-book and solar cell – I hope you, the reader, will appreciate how $E = hf$ also changed the world. Indeed, in honour of Bill Bryson, I did consider calling this book *A Short History of Nearly Everything Else*.

Many other science books explore quantum ideas. I have suggested some in the Bibliography. Usually they describe esoteric features like the paradox of Schrödinger's cat rather than how silicon chips work. I aim to show you that the more practical, everyday applications of quantum ideas can be just as fascinating.

Also, understanding quantum ideas will help you to see how similar silicon chips are to the solar cells that provide us with solar electricity. We can all appreciate how silicon chips have revolutionised our lives. Solar cells, which are often referred to as photovoltaics (or PV for short), are playing an important part in the solar revolution.

As the history of quantum ideas unfolds in Part I, we will also discover how the application of $E = mc^2$ slowed down the development of some of the more peaceful uses of $E = hf$.

Understanding how renewable technologies work will be important in Part II when we will meet the amazing range of solar technologies that can help us reduce our carbon footprint. It will also help you appreciate how complementary the solar technologies are in supplying our electricity.

We will also need some physics in Part II to counter the arguments of those who oppose the solar revolution. Many people find the debate about energy options extremely confusing. Here is a typical argument from two commentators, Ted Nordhaus and Michael Shellenberger, in *The Wall Street Journal* on 22 May 2013. They are clearly sceptical about renewable energy and critical of two solar supporters, Robert F. Kennedy Jr of the Natural Resources Defense Council and Bill McKibben, who had been writing in *The Daily Beast*. You can find the reference to this, and other quotations, in the Bibliography. The *Wall Street Journal* pair are criticising the solar supporters who had been praising the German achievements in renewable energy:

> Messrs. McKibben and Kennedy, for instance, have boasted that on one day in 2012 half of Germany's electricity came from solar … The real story is much more sobering. In 2012, solar generated less than 5% of Germany's electricity.

So are Messrs McKibben and Kennedy exaggerating German achievements by an intemperate factor of ten? Should the solar supporters have said 5 per cent rather than 50 per cent? No – these numbers refer to different quantities! Nordhaus and Shellenberger are confusing electrical *power* and electrical *energy*. I am guessing this was unintentional as it happens very often in energy debates. When we look at Germany's achievements in renewable energy in Part II you will find that what Nordhaus and Shellenberger should have said was:

> … on one day in 2012 *around noon*, half (50%) of Germany's electrical *power* came from solar … In 2012, solar generated less than 5% of Germany's electrical *energy*.

We will later find why solar power and solar energy differ by such a large factor. We will also understand the impact solar achieved while contributing less than 5 per cent of the electrical energy in Germany: a significant reduction in the wholesale price of electricity on the grid.

INTRODUCTION

Energy and power not the same thing, but they are related. There are many types of energy that can be transformed from one form to another. Power transforms energy. In fact, power is the amount by which energy changes in a second.

Interestingly, I find that one type of commentator has no problem at all in distinguishing energy and power. Football commentators describe my grandsons' favourite player, the Welshman Gareth Bale, as running very fast with the ball and showing lots of *energy*. This is correct; the faster he runs, the higher his energy of motion. When commentators talk about how powerfully Bale hits a dead ball they are also correct. Getting the ball to move fast and accurately is all about his skill in applying *power* at the instant his foot is in contact with the ball. That way Bale can change the energy of the ball from zero to a very high energy of motion as it hits the back of the net. Power is more important than energy as it determines when energy changes and by how much.

Another way to remember the distinction is that the power provided by the fusion reactor at the centre of the sun has been essentially constant for more than four billion years and is likely to continue to remain so for another four billion years. The power of the sun turns nuclear energy from its core into the energy of sunlight. I will share with you the excitement of the struggle that physicists had, through three centuries and two false starts, to explain the nature of the energy in sunlight. Their confusion was resolved in 1905 by Einstein's interpretation of $E = hf$.

If you hold your hand in the sun you can feel the power of sunlight. Strictly solar power is the amount of sunlight energy hitting your hand in one second. Solar power converts the energy in sunlight into heat energy, so your hand feels warmer.

Most importantly we will find many ways in which the energy in sunlight can be turned into useful heat energy and electrical energy. In particular, in solar cells the energy in sunlight can be simply converted into electric energy, which can supply human-kind's power requirements many times over. We will meet other renewable technologies that you may not have been aware originate in solar power.

There is a simple way to check if numbers you come across in the energy debate are referring to energy or power. If the number represents *power*, then the number must be followed by a letter W. This stands for watts, the unit of power, named after the Scottish engineer James Watt who developed one of the earliest steam engines. You will often find a k, M or G before the W – as in kW, MW and GW – where the k, M and G stand for a thousand, a million and a thousand million respectively.

If the number represents *energy*, a number of units are commonly used. The most likely one in any debate about electricity supply is kWh, which is short for 1,000 watts *times* hours. A small electric kettle of 2 kW power which boils water for half an hour will have consumed 1 kWh of electrical energy.

Should a commentator present a number for an amount of 'electricity' without W or Wh, feel free to ignore them.

In the third part of this book, as well as setting out my manifesto and giving you suggestions on how to join the solar revolution, I will also speculate on how this revolution might develop in the future. I hope it does succeed in averting the three existential threats that face us. I think it is the only chance we have.

Why should you have any confidence in my opinions? My views on the dangers of nuclear waste were developed from more than a decade of (spare time) research into plutonium production in civil reactors. This study was in turn informed by two decades spent researching experimental particle physics at the University of Birmingham, CERN, the University of California Berkeley and Imperial College London.

My views on the future of photovoltaics are based on two decades of research at Imperial leading a group that developed what we called QuantaSol technology. In April 2013 the US company JDSU announced they had manufactured triple junction cells with 42.5 per cent efficiency for conversion of solar power to electrical power using QuantaSol technology in two of the three subcells. This was at the time the world's highest efficiency for production solar cells and around three times the efficiency of current rooftop technology. I can reassure you that, as neither of the solar cell companies that I have co-founded now exists, my predictions are not influenced by any current financial involvement. Though, if this technology does eventually take off, I and my colleagues may benefit from some patents.

I hope this book will equip you with the information you need to understand the answer to the burning question of our age: how to supply the power our society demands while avoiding environmental catastrophe and nuclear disaster. I hope you will then play your part in promoting the solar manifesto and join the solar revolution. This, I firmly believe, is the only way our descendants can continue to enjoy the power of our sun for the four billion years it, hopefully, will continue to burn.

I

The History of the Semiconductor Revolution

ONE

We are Stardust

We are stardust
Come from billion year old carbon
We are golden
Caught up in some devil's bargain
And we've got to get ourselves
Back to the garden

Joni Mitchell, 'Woodstock'

Joni Mitchell has her physics right; we are all stardust. To explain what she is singing about, let's first travel back to a few minutes after the big bang when our universe was born.

Joni's classic song suggests that if we consider what we are made of and where we came from, we will find pointers to where we should go. So, what are we made of?

The human body, and indeed all the matter in the universe, is made up of *atoms*. There are 92 types of stable atom in the universe. Each different type of atom, known as an *element*, has a different number of *electrons*. The electron is one of the tiny heroes of our story. Electrons orbit the very much heavier *nucleus* at the centre of the atom, rather like planets orbiting the sun. *Hydrogen*, the lightest atom, has one electron and the simplest nucleus; it consists of a single heavy particle called the *proton*. *Helium* has two electrons and a nucleus containing two protons. The more electrons there are in the atom, the more protons there are in the nucleus and the heavier the atom. *Uranium*, which is the heaviest stable element, has 92 electrons orbiting the 92 protons in its nucleus. The proton is nearly 2,000 times heavier than the electron. Hence most of the weight of our body is due to the nuclei of the atoms.

Where do the nuclei, which make up most of our bodyweight, come from?

A brief history of the first ten billion years of our universe

Astrophysicists date the big bang at more than 14 billion years ago (a billion is a thousand million; it is a number which will feature a lot in our story). The complex physics of what happened in the first seconds after the big bang, though absolutely fascinating, is not relevant to our story here. Suffice to say that, three minutes after the big bang, all the matter in our universe was dominated by the very electrons and protons that form our bodies and our earth today.

How come we are made up of stardust? A mere 140,000 years after the big bang the temperature of the universe had cooled sufficiently for electrons and protons to condense to form hydrogen atoms. Thereafter, clouds of hydrogen gas coalesced into galaxies, which condensed further to form stars like our sun. The force of gravity pulled the hydrogen atoms closer and closer together, and the temperature and density of the hydrogen at the core of these primitive stars increased. As the temperature and density rose higher, the colliding protons, which were the nuclei of the hydrogen atoms, started to fuse together to form helium nuclei. From this moment on, solar power became the dominant energy source in the cosmos.

The burning of hydrogen to form helium nuclei releases vast amounts of energy. Much of this energy is in the form of sunlight. The pressure of the light forcing its way out of the star balances the inward pressure of the hydrogen falling into the core. This balance can keep stars stable for billions of years.

Towards the end of the life of a star, the hydrogen fuel starts to run out and less sunlight is produced. More hydrogen falls into the star than the escaping sunlight can balance and the density at the core of the star increases further. Eventually the temperature and density become sufficiently high that colliding helium atoms start to fuse to form heavier nuclei. This process continues to generate the nuclei of heavier elements such as *carbon*, without which life would never have developed. Most stars end their lives with some form of explosion, so these heavier nuclei litter the cosmos.

Nuclei with up to 26 protons were generated in this way. Heavier nuclei were generated in the even more extreme conditions found when some stars end their lives in a catastrophic explosion known as a *supernova*. In these cases, the light emitted by a dying star can briefly outshine the whole of its galaxy. The extraordinary conditions that occur in the course of these explosions allow nuclei to fuse together to form the heavier nuclei.

Our sun, the great hero of our story, is a fairly typical star. Most importantly for life on earth, it is a *second-generation* star. The sun and its planets are relatively young on the cosmological timescale – only around four-and-a-half billion years old. Many first-generation stars had already ended their lives by the time our sun formed. By then, the nuclei produced by first-generation stars and supernovae littered the galaxy. During the early years of its life, our sun dutifully swept up this waste to form our solar system. I find this idea mind-boggling, but astrophysicists insist that is what happened. The nuclei of every atom heavier than helium, in our bodies and in the ground beneath us, came from the death of another star.

So Joni Mitchell is correct: we mostly consist of the billions of years old waste of burnt out stars. Every nucleus of every carbon atom in your body was created in the death-throes of a first-generation star. We are, quite literally, formed of '*billion year old carbon*'.

The period in which the first-generation stars were living and then burning out is known as the period of *nucleosynthesis*. It was the period in which nuclear physics reigned supreme. Did nucleosynthesis stop with the nucleus of the element uranium? It did not. Many heavier nuclei were produced, including the nucleus of plutonium. An atom of plutonium has 94 electrons and 94 protons.

Plutonium is one of the villains of our story. It has many properties that are uniquely dangerous. If a nuclear worker inhaled one millionth of a gram of plutonium this would be above the maximum permissible safety level. Had significant amounts of the plutonium produced by first-generation stars survived, life as we know it could not have developed. Fortunately, all nuclei heavier than the uranium nucleus are *unstable*. That is, they decay to lighter nuclei in a process known as *radioactivity*. We will find that the lifetime of the plutonium nucleus is very long on the human life-scale, but on the cosmological timescale it is quite short. Hence all the plutonium had decayed away well before life began on earth.

In later chapters, we will learn how, in the traumatic circumstances of the Second World War, plutonium was re-created in our solar system. It is appropriately named after Pluto, the god of the underworld. It can be used to produce massive destruction and to produce electricity. Perhaps Joni's '*devil's bargain*' refers to plutonium.

But what did she mean by '*And we've got to get ourselves back to the garden*'? Four-and-a-half billion years ago our earth started forming from the nuclear waste of the first-generation stars. For the first two billion years the earth was solidifying and the seas, dry land and a primitive atmosphere were forming. Then sunlight, which is generated as a by-product of the fusion of hydrogen nuclei to helium nuclei in the sun's core, started driving an amazing new process in the oceans: the creation of life. I regard this process as the greatest achievement of nuclear fusion. I also think the core of the sun is the best location for a nuclear fusion reactor.

For the last two-and-a-half billion years of the four-and-a-half billion year lifetime of our earth, sunlight has been driving the increasingly complicated chemical reactions known as *photosynthesis*: plants take carbon dioxide out of the air to produce the carbon-based compounds that make the plant grow. The oxygen we breathe is a waste product. The creation of life is driven by the power of sunlight.

There are important differences between the eight billion years or so of cosmological nucleosynthesis, which was driven by nuclear fusion, and the two-and-a-half billion years of terrestrial photosynthesis on earth, which is driven by sunlight. First, I will explain what I mean by *burning*. It is the generation of heat and light energy by consuming a fuel while producing waste. Uncontrolled burning can also result in an explosion. Not all burning is bad news. Nuclear fusion in stars is a burning, and the waste products are elements like carbon from which life is formed. The waste products of the catastrophic uncontrolled burning at the end of a supernova's life are our heaviest nuclei. One by-product of the burning of hydrogen and helium in the sun is the sunlight which has driven the evolution of life on earth.

Photosynthesis is not burning. It is a creative, chemical process rather than the destruction of a fuel. Energy, in the form of sunlight, is absorbed and stored in the plant. Complex carbon-based compounds are created from carbon dioxide and water while liberating the oxygen vital for the development of life on earth. I interpret Joni Mitchell's plea that '*we've got to get ourselves back to the garden*' as meaning that humankind has to get back to the situation when sunlight provided all our primary energy needs. This will become clearer as we look in more detail at the important events in the most recent billion years of the earth's history.

A brief history of the last billion years on earth

It is not easy to grasp what a billion (a thousand million) means. I think it helps to appreciate how long a billion years takes if we shrink it down to just one normal calendar year. This means we are considering time passing at a rate of one second equivalent to about 32 years, which, conveniently, is approximately a generation. A billion will turn up a number of times in our story, so it is helpful to remember there are approximately a billion seconds in a generation.

At one generation a second you should be prepared for a roller-coaster ride towards the end of December, but early in the year progress is much more sedate. One billion years ago, on 1 January in our equivalent year, evolution was happening slowly, even at the rate of one human generation a second. The only form of life on earth was in the oceans. Algae-like, multi-cell organisms had taken over from the primitive cells, which had started the task of photosynthesis a-billion-and-a-half years earlier. They were taking carbon dioxide and water and using the power of sunlight to convert them into the material they needed to grow. The waste product was oxygen, which would eventually provide the earth with an atmosphere that could support life.

As more and more life forms developed under the sea, more oxygen was formed until, around 17 June, the pace of evolution started to speed up, in what is known as the *Cambrian explosion*. Many diverse species developed in the sea in a relatively short period. Experts disagree about how long. It took between three days and three weeks on our scale. Then came two more very significant events – the first early in July when plant life spread from sea to shore, followed in August by fish as they evolved into amphibians.

Late August to mid-September was an important time for our story. It is known as the *Carboniferous period*. Lush vegetation covered much of the land mass. The decayed remnants of these plants eventually ended up underground, transformed into today's coal. Around 1 October a cataclysmic event occurred – the biggest of the mass extinctions. It was probably triggered by a large meteorite hitting the earth. Some life forms survived and evolution subsequently produced the dinosaurs, which dominated from mid-October to early December. In this period, much of the decayed vegetation and algae in the sea ended up as sediment that eventually formed today's oil reserves. On 7 December another mass extinction occurred. This time it was almost certainly caused by a meteorite, which left a large crater in the sea off the Yucatan peninsula in Mexico.

Some early mammalian life forms survived this extinction and evolved relatively rapidly thereafter. Sometime between noon on 29 December and the early hours of 30 December our human ancestors became a species distinct from chimpanzees.

It was not until 31 December, in our representative year, that human development really started to speed up. Late in the morning on the last day of the year our upright walking ancestors in Africa and Asia, known as the Ergasts, discovered fire and the burning began. At two minutes to midnight on New Year's Eve, Stonehenge was built. Around nine seconds before midnight, the first coal-burning steam engines started working. At three seconds before midnight, the motor car was invented, the large-scale exploitation of oil started and the two equations $E = mc^2$ and $E = hf$ were discovered. Around two seconds before midnight, plutonium was reinvented.

It is midnight on 31 December. What does the future hold? Clearly in much less than a second we must make severe cuts in our carbon emissions if runaway global warming, with its profound effects for life on earth, is to be avoided. It is likely that in less than three seconds both the exploitable oil and uranium reserves will be depleted. By 3 a.m. on New Year's Day the separated plutonium from the first two generations of civil nuclear reactors in the UK will have decayed to a level at which it will no longer be of interest to terrorists. That may not sound long, but in human terms that is more than 10,000 generations away.

What can we learn from history?

We can draw some important conclusions from this brief history. For almost all cosmological time, apart from the last nine generations or so, solar power has been the primary resource driving the evolution of our universe, first, in generating the nuclei of atoms and then, for the last two-and-a half billion years, powering photosynthesis to generate life.

Second, in just over three generations, we have used approximately half the exploitable oil, which it took photosynthesis followed by underground pressure around eight million generations to produce. The depletion of this important resource was a factor in two Gulf wars. The majority of scientists agree that the carbon dioxide generated by burning this oil, and the coal produced in the Carboniferous period, is a major contribution to global warming.

Third, in about two generations, we have used approximately half the exploitable uranium it took solar power 250 million generations to produce. We must keep the plutonium waste produced in two generations out of the environment for well over 10,000 generations and the nuclear industry does not yet know how to do this or where to store it. The problems of nuclear waste were highlighted by the nuclear disaster at Fukushima in Japan in 2011. Around 150,000 people have been evacuated, many may never be allowed to return. One problem for the brave technicians who were forced to try to contain the radiation is that many years of irradiated waste was stored in cooling ponds inside the reactor buildings because there was *nowhere else for the waste to go*.

There are other problems with nuclear waste. If I were to show you a picture of circles of 4,000-year-old stone blocks on Salisbury Plain, you would no doubt recognise Stonehenge. But do you know why it was built? What if a more advanced alien civilisation had stored its nuclear waste deep under Salisbury Plain and left the stones to warn us? Perhaps the Department for Environment has had the same idea? There have been plans around for many years to dig a tunnel to take the nearby trunk road underground to help preserve the stones, but there has yet to be any action.

We don't know if Stonehenge is saying that nuclear waste is stored there. How are we going to alert future generations to the contents of our civil nuclear waste store when it is eventually built? To do so, we must eventually ensure that plutonium alone must be kept out of the environment and out of terrorist hands for *more than 80 times* as long as Stonehenge has stood on Salisbury Plain. What language, what symbols, what computer memory will last that long?

If humankind is to have a long-term future on earth, we must stop burning fossil and nuclear fuels and ensure we return to using the sun as our primary source of power – as it has been for the billions of years of evolutionary history of life on earth. As Joni Mitchell says, '*And we've got to get ourselves back to the garden*'.

We can get back to the garden. Our sun *can* provide all our primary energy needs again. The sunlight falling on the earth in *one hour* is more than enough to supply all the energy demands of humankind for *one year*. As we will see, there are solar technologies that are ready and waiting to take on this challenge.

Astrophysicists believe the sun is about halfway through its lifespan. Restoring sunlight to its rightful position as our primary power resource is the best way to ensure humans will be around to benefit from the sunlight the sun will produce over the next four billion years or so.

But what is this sunlight, which has driven evolution on earth? If we wish to harvest it, we need to know what it is made from. In the next two chapters we will look in more detail at another roller-coaster history – the story of physicists' pictures of sunlight and how dramatically they changed in just three generations.

TWO

What is Light?

Annual income twenty pounds, annual expenditure nineteen nineteen and six, result happiness. Annual income twenty pounds, annual expenditure twenty pounds ought and six, result misery.

Charles Dickens, *David Copperfield*

For two-and-a-half billion years sunlight has powered the evolution
of life on earth. If humankind is to take advantage of the four
billion years or so that the sun will continue to shine, we need
a solar revolution. Understanding the nature of sunlight was the
first step in this revolution.

Two important changes occurred in the way physicists picture
sunlight in the nineteenth century. The first change occurred right
at the start of the century. The second change in 1865 was truly
revolutionary. It led 130 years later to the mobile phone. In between
these changes, a breakthrough came in an apparently unrelated part
of physics. In 1831 a practical way was found to generate electricity.
Unexpectedly, the 1831 discovery proved crucial to the 1865 revolution.

In this chapter we will explore these important developments,
which all took place in nineteenth-century Britain. The dramatic
opening ceremony of the 2012 Olympics reminded a world audience how Britain led the Industrial Revolution by covering much
of our green and pleasant land with coal-burning chimneys. The
harshness of daily life for working people was vividly captured in
Charles Dickens's novels. Mr Micawber's famous advice encapsulates
the monetary knife-edge on which many people lived. I put my own

financial challenges in context by reminding myself that my maternal grandfather was born in a Victorian workhouse.

While the masses struggled to make ends meet in the nineteenth century, a small number of physicists and engineers were making major discoveries that have made our lives so much easier. These started at the turn of the century, with the appointment of Thomas Young to a professorship at London's Royal Institution. He was an English gentleman, a polymath and, like a number of famous physicists at the time, a man of independent means. At the Royal Institution he resolved a major conflict between those physicists who thought sunlight was a particle, like a tiny billiard ball, and those who believed sunlight was a wave. By passing sunlight through two closely spaced parallel slits he formed a pattern on a screen, which was only explicable if sunlight consisted of waves. This major physics discovery was achieved while Young was a full-time medical practitioner with a surgery a few blocks from the Royal Institution. He even found time to study Egyptian hieroglyphs and later helped to decode the Rosetta stone.

One of Young's successors at the Royal Institution, Michael Faraday, had a very different background. A bookbinder's apprentice for seven years, he started work at the Royal Institution as a secretary and valet to one of the professors. On a trip around the top scientific laboratories of Europe, Faraday was made to take a seat outside the coach as he was the son of a blacksmith. Faraday's greatest achievement was the law of 1831 that bears his name. Nowadays, we generate most of our electricity using Faraday's law. The exception is solar photovoltaics, the new technology that is fundamental to our story.

If sunlight is waves, what is waving?

Young's two slit experiment showed that sunlight consisted of waves – regularly repeating motions of peaks and troughs, like waves on the surface of water. The distance between two successive peaks is called the *wavelength* of the wave. Our eyes see beauty in the rainbow because each colour has a different wavelength. Blue sunlight on the inside of the rainbow has about half the wavelength of red sunlight at the outside.

But what is actually waving? Is sunlight like waves on the surface of water? Or is it like sound waves? Water waves are easy to picture, but sound waves, travelling through the air, less so. Our atmosphere consists of molecules (pairs of oxygen atoms and nitrogen atoms) that are moving around fast in all directions. Though air molecules are very small and very far apart, they move around at such high speeds that they continually bounce off each other. Hence, they do not actually get very far. A typical air molecule bounces off other air molecules over a billion times a second.

When a cluster of air molecules is knocked by a vibrating violin string, the molecules move away from the string and bounce off more distant air molecules. The original molecules rebound back towards the violin ready for more punishment when the vibrating string returns. The molecules that have been struck by the original ones bounce off other air molecules further from the violin. In this way a wave of rebounding air molecules carries the original vibrations of the violin to your ear.

When the wave reaches your ear the air molecules rebound from your eardrum. That is how you are able to hear the sound of the violin. It is actually the back-and-forth wave motion of the air molecules that is responsible for your hearing the violin; so you would not be able to hear music on the airless moon.

Air molecules carrying a sound wave behave like dancers at a crowded rave. If a reveller has drunk too much and gyrates madly, nearby dancers will try to avoid his wild thrashings and in doing so will bump into their neighbours who are further away. The disturbances caused by the drunk spread across the dance floor as waves, while everyone keeps on dancing.

WHAT IS LIGHT?

It was the nineteenth-century Scottish physicist James Clerk Maxwell who discovered what is waving in sunlight. Maxwell was born into a land-owning family in south-west Scotland. His mother died of cancer when James was only eight. She was 48; the same age as her son when he died 40 years later of the same disease. Though Maxwell is best known for his theoretical work at Aberdeen and Edinburgh Universities and Kings College London, in later life he became the first Cavendish Professor of Experimental Physics in Cambridge.

The moment in 1865 when Maxwell realised that sunlight was a very different wave motion to either water waves or sound waves must have been one of the greatest highs that any scientist has experienced. It is a story rarely told in popular science books and is all the more interesting as he stumbled on the explanation by accident.

At the time he made his discovery, Maxwell was not even thinking about sunlight. He was trying to understand what happens when a loop of wire carrying an electric current is cut so that the current stops. To explain why the current does not stop immediately, Maxwell had to make a small change to an equation first proposed by the French physicist André-Marie Ampère. While thinking about his new equation, Maxwell stumbled upon the possibility that electric and magnetic fields could form a self-sustaining wave. He realised almost instantly that these new waves were sunlight.

Most scientists occasionally have a new idea that looks promising. We then undergo mental agonies while colleagues with the necessary expertise or a student operating the relevant equipment perform the experiment that confirms the idea is publishable or consigns it to the waste bin of scientific history. Maxwell did not have to undergo such agonies. The experiments that confirmed his new picture of sunlight had already been made.

For a short period of history Maxwell was the only person who knew that sunlight is made from waves of time-varying electric and magnetic fields. Did he tell Mrs Katherine Maxwell? 'You will be the only woman on earth, my dear, who knows what sunlight is.' We do know he told a maths teacher cousin. A letter written in January 1865 contains some typical Scottish caution: 'I have also a paper afloat, containing an electromagnetic theory of light, which, till I am convinced to the contrary, I hold to be great guns.'

Maxwell discovered that sunlight is a much more sophisticated wave motion than sound waves. The latter, remember, are the regular bashings of air molecules, superimposed on their random motion, like our dancers at a rave. The picture of sunlight provided by Maxwell is of waves of electric and magnetic fields, varying in time and space like two perfectly choreographed dancers in a classical ballet. You are reading these words because sunlight or lamplight has waved its way from the page or screen to your retina at the speed of light. Loops of electric field have generated loops of magnetic field, which in turn have generated loops of electric field untold billions of times en route from page to brain. At all times, these dancing partners – the electric and magnetic fields – are following the strictly choreographed mathematics of Maxwell's equations.

If you follow through my explanation of electric and magnetic fields you will not only understand how most of our electricity is generated, you will also be able to experience something of Maxwell's excitement when he made his revolutionary discovery. It has made our lives so much easier than those of our Victorian ancestors.

What are electric fields and magnetic fields?

The electric field is responsible for the forces between electric charges. You may remember the mantra from school science: like charges repel, unlike charges attract. Well, the electric field causes the force of repulsion between two like charges and the force of attraction between two unlike charges.

All the various particles that make up matter have either *positive* charge (like a proton), negative charge (like an electron) or have no charge at all, in which case we say they are *neutral*. We will later meet another of the villains of our story, the *neutron,* which, as its name implies, is neutral.

Atoms consist of a number of electrons orbiting a very small heavy nucleus containing an equal number of protons. All electrons have exactly the same negative electric charge as each other. All protons have exactly the same positive charge as each other. Since an atom is exactly neutral when it has equal numbers of electrons and protons, the charge on one electron must be the same as the charge on one proton. The only difference is that one charge is a positive number and the other charge is a negative number. Physicists have chosen the proton to have a positive charge. Like poets they favour alliteration.

It is an amazing coincidence that every electron has exactly the same size of charge as every proton. Fortunately the coincidence is exact; otherwise the atoms in your body would repel each other and you would, quite literally, fall to pieces.

The force of attraction between the unlike charges of the electron and proton holds the electrons in orbit round the nucleus and keeps our atoms stable. This is very much like the *gravitational* force of attraction of the sun on the planets holding the solar system together.

It was the blacksmith's son Faraday who first introduced the idea of the electric field. It is something that extends from a positive charge through space to a negative charge and is responsible for the attractive force between them. An electric field also extends from a positive charge through space to another positive charge and is responsible for the repulsive force between them. Electric fields are also responsible for the repulsive force between two negative charges.

Physicists measure the electric field at a particular position by placing a small positive charge at that point. The direction in which the positive charge is pushed is the direction of the field.

If the small positive charge is replaced by a small negative charge it will be pulled by the same field in the opposite direction.

The magnetic field is a sister field to the electric field with many similarities. It is something that extends through space from a north magnetic pole to a south magnetic pole. Physicists measure the magnetic field at a particular position by placing a small magnet at that place. It will then point along the magnetic field.

You have probably seen the magnetic field around a bar magnet demonstrated in a school science lesson. Iron filings are scattered around a bar magnet laid on a piece of paper. The iron filings act like small magnets and align themselves along the lines of the magnetic field.

In the Bibliography there is an internet reference where you can look at the pattern of the magnetic field revealed by the filings. The lines of the magnetic field spread out from the north pole at one end of the magnet, loop around and converge towards the south pole at the other end.

Another way to produce a magnetic field was discovered by the Frenchman André-Marie Ampère. His father was a successful businessman from the region around Lyon. His son excelled in mathematics and rose to become professor of mathematics at the École Polytechnique in Paris. His important contribution to physics was an observation which we now interpret as due to a magnetic field appearing around any wire in which an electric current is flowing.

The magnetic field forms a circular loop round a long straight wire carrying an electric current. This can be shown by small compasses placed on a board through which a wire passes. There is a picture of this experiment too at an internet address in the Bibliography. The compass needles all point round the circumference of a circle which is centred on the wire.

The Bibliography also provides the link to a demonstration on the web. You see the magnets all pointing north when there is no current in the wire. If you move your mouse to switch on the current the compass needles move until they form a loop around the wire. You can reverse the direction of the current in the wire and the compass needles point the opposite way.

Faraday's new source of electricity

Faraday's greatest contribution was to discover a new and more practical way to generate electrical power than the early batteries in use before 1831. We now use Faraday's method to generate nearly all our electricity whether from fossil or nuclear fuels or renewable sources. As we will later find, only solar cells do it differently.

To understand Faraday's achievement, first think about what it takes to boil an electric kettle. An electrical engineer says the electrical power you need depends on two things: the *voltage* and the *current*. The electricity company provides you with a voltage between two contacts in the wall socket. The voltage drives electric current along one wire in the flex to boil the kettle. The current then flows back down a second wire in the flex to the other contact. The power measured in watts is found by multiplying the voltage measured in volts by the current measured in amps.

A physicist sees electrical power differently. Following Faraday's discovery of electric fields, a physicist pictures such a field running from one contact in the socket down the first wire, though the kettle and back down the second wire to the other contact. Faraday pictured electric current as this field *pushing* positive charges through the wire. Nearly seventy years later in 1897, J.J. Thomson, Cavendish Professor of Experimental Physics at the University of Cambridge, showed that electric currents were due to electric fields *pulling* negatively charged electrons through the wire. Exactly 50 years further on, in 1947, some American physicists showed that useful electric currents in semiconductor crystals could be produced by electric fields pushing positive charges through the crystal. This discovery was crucial to the development of modern personal computers. So you could say that Faraday was well ahead of his time.

To complete the physicists' picture of making a cup of coffee, we now think of the electric field pulling the negatively charged electrons in the second wire through the element in the kettle. The electrons bash into the atoms of the element making these atoms move back and forth faster. The faster the atoms in the element move, the hotter the element gets and eventually the water boils.

Faraday was intrigued that Ampère had shown that an electric current in a wire generates a magnetic field. Could a magnetic field produce an electric current, he wondered? He tested this idea with a series of experiments. At first the idea did not seem promising. A magnetic field passing through a loop of wire does not produce an electric current in the wire. He continued his experiments and, in 1831, crucially and unexpectedly, Faraday observed an electric current in the loop when he *moved* a magnet relative to the loop of wire. A magnetic field through a loop of wire *does* produce a current in the wire but only if the magnetic field is not steady but *changes with time*. The electric current is produced by an electric field in the loop of wire, which is related to the way the magnetic field changes by an equation known as Faraday's law.

The pace of technological advance was slower in Victorian times than today. It took a full 50 years before this immensely important discovery was exploited commercially. We will find that the positive charges that the Americans discovered in 1947 were commercialised well within the next decade.

Maxwell's equations

In 1865 Maxwell discovered that sunlight consisted of waves of electric and magnetic fields and found a way to describe their behaviour exactly in mathematical equations, one of which was Faraday's law.

There are plenty of popular books about quantum theory, but few about Maxwell's equations. Stephen Hawking's *A Brief History of Time* and Peter Atkins's *Galileo's Finger: The Ten Great Ideas of Science* each devote only one paragraph to Maxwell's theory. Yet Maxwell's equations and quantum ideas together underpin all the technological breakthroughs of the semiconductor revolution. The silicon chip, laptop computers and mobile phones all depend for their operation on Maxwell's equations and quantum theory.

Perhaps popular physics authors so rarely explain Maxwell's revolutionary picture of the waves that make up sunlight because four complicated mathematical equations are needed to explain his theory. In *A Brief History of Time*, Stephen Hawking says that the book only contains one equation, $E = mc^2$, because 'Someone told me that each equation I included in the book would halve the sales'. By this reckoning, including four equations in this chapter could reduce sales to one sixteenth of the book's potential. But I think it is worth the risk because Maxwell's revolutionary picture is nearly as mind-blowing as quantum theory. Also, his equations explain how your mobile phone works.

To make these four equations feel less daunting, I will describe them in words, although I will allow myself one new equation in mathematical symbols, $E = hf$.

A physics equation sets out the mathematical relationship between measurable quantities. Mr Micawber's advice 'Annual income twenty pounds, annual expenditure nineteen pounds nineteen and six, result happiness. Annual income twenty pounds, annual expenditure twenty pounds ought and six, result misery' is, in fact, two equations in words. *David Copperfield* was written 15 years before Maxwell's discovery. We can all measure the quantities in Dickens's equations. If you are British and of a certain age, you may remember that there were 12 old pence in a shilling and 20 shillings in a pound. So Dickens's quotation consists of two equations in words:

THE HISTORY OF THE SEMICONDUCTOR REVOLUTION

Plus sixpence = Happiness
Minus sixpence = Misery

Micawber's equations relate home economics to our well-being. Maxwell's four equations explain the relationships between electric charge, electric fields and magnetic fields.

Dickens's equation describes what happens to a bank balance with *time*. Your bank balance falling from one day to the next is bad news. But your financial worth can also change with *position* if money is transferred from your current account to a building society account.

Maxwell's first two equations, which say how electric fields and magnetic fields change with position, were borrowed from a German physicist Carl Friedrich Gauss. Like Faraday, Gauss came from a poor family. His father was a gardener in Brunswick, Germany.

Gauss's first equation explains how an electric field varies with position. Electric fields diverge away from a positive charge and converge towards negative charges. This is just like the magnetic field diverging from a N pole and converging towards a S pole. Gauss's second equation says that magnetic fields always form closed loops. This is very clear in Ampère's experiment. The loops can also be seen from the iron filings if you ignore the filings on the magnet itself.

Maxwell's third equation was the mathematical expression we call Faraday's law. Remember 34 years earlier Faraday had shown that if a magnetic field passing through a loop of wire varies with time, there will be an electric current in the loop. Remember also that Faraday pictured the current in the loop of wire as being due to an electric field in the wire pushing or pulling charges. Faraday's law in words says:

Electric field loop = change in time of magnetic field through loop

Maxwell's fourth equation was also borrowed. In this case, it was from Ampère, who showed that loops of magnetic field appeared round the electric current flowing in a wire. The mathematical equation that summarises Ampère's law can be expressed in words as:

$$\text{Magnetic field loop} = \text{electric current in the wire}$$

So the Scot, Maxwell, borrowed two equations from the German Gauss, one from the Englishman Faraday and the fourth from the Frenchman Ampère. This was a remarkable early example of European scientific collaboration and, indeed, co-operation between social classes. So why do we call all four Maxwell's equations?

Maxwell's magic moment

In 1865, Maxwell made a small, but beautifully formed, addition to Ampère's law. The results were revolutionary.

Maxwell was not trying to find out what sunlight was made from, but thinking about something more prosaic. If you have an LED (light emitting diode) monitor on your laptop charger or flat screen TV you will probably have noticed that, when you pull the plug from the wall socket, the LED light does not immediately turn off, despite the fact that there cannot be any current flowing from the mains. Maxwell had absolutely no idea that 150 years later, first, such devices would exist and, second, they would depend for their operation on the amazing fact he was about to stumble upon.

Maxwell's scientific curiosity had been aroused by noting that Ampère's law does not explain why the current in an electric circuit fades away when a switch is thrown. Ampère's law is very clear. Immediately there is no electric current in a wire, there will be no loops of magnetic field around the wire. When the plug is pulled and there is no current, Ampère's equation in words becomes:

$$\text{Magnetic field loop} = 0$$

Maxwell reasoned that this equation could not be correct when the circuit was broken because the current clearly took time to fall to zero. Perhaps something was missing from Ampère's law? Perhaps a small extra, contribution that was only important when the circuit was broken? But what expression should be used for this extra contribution?

Maxwell's argument for the missing contribution is rather subtle. I think it is easier to explain his choice in another situation where the electric current is zero. Think about a region of outer space where there is nothing, just a vacuum. No solid matter, no electrons, no protons, no electric currents, just empty space. Faraday's law and Ampère's law should still apply in outer space. In our chosen region of empty space, they become:

$$\text{Electric field loop} = \text{change in time of magnetic field through loop}$$
$$\text{Magnetic field loop} = 0$$

The right-hand side of Ampère's law is again zero. There are no electrons or protons in this region of empty space and certainly no moving charges; hence there are no currents. Look carefully at these two equations and imagine you are Maxwell thinking that something might be missing from the second one. Don't these two laws look asymmetric? One could make an addition to Ampère's law, which would make the two equations symmetric:

$$\text{Electric field loop} = \text{change in time of magnetic field through loop}$$
$$\text{Magnetic field loop} = \text{change in time of electric field through loop}$$

WHAT IS LIGHT?

Isn't Maxwell's new equation rather more symmetric with Faraday's law?

Maxwell's discovery of the missing term was one of the first examples of the use of symmetry in physics. Nowadays, symmetry is one of the most important tools in the armoury of a theoretical physicist. A modern-day Maxwell would use a symmetry argument in preference to an argument based on breaking a circuit. Indeed, some physicists believe that eventually we will discover single magnetic poles, which are the equivalent of the electric charges of electrons and protons. Then Maxwell's equations would be completely symmetrical.

To a physicist, a beautiful theory is one that is both elegant and symmetrical; but it must also correctly predict the results of experiment. How can we test if Maxwell's discovery is correct? Most theories are first tested with a 'thought experiment'. Here's one suggestion:

Think about doing an experiment in a region of empty space. The two equations in words above, Faraday's law and Maxwell's new equation apply in this region. Image a single electron is brought very close to this region of empty space but remains just outside it. There will be an electric field, which will extend from the electron into the empty space like the magnetic field around a S pole. But there was no electric field in the empty space before we brought up the electron. It is important to realise that bringing up the electron near to the empty space *makes the electric field inside the region change with time*. If Maxwell's new equation is correct, this change of electric field with time will produce loops of magnetic field in the empty region.

Now comes the really interesting bit. These loops of magnetic field were not there before we moved the electron. So we also have a new magnetic field in the empty region, which has changed with time. But hang on! Doesn't Faraday's law say that if there is a change in time of a magnetic field there must be loops of electric field around it? This is a new electric field that was not there before. This change in time of this new electric field will in turn produce new loops of magnetic field that were not there before and that means ...

Is your head going round in circles? Maxwell's may well have been the first time he thought about this problem. We are going from his new equation to Faraday's law and back again. This is all because of the new contribution which Maxwell had added to Ampère's law. Simply by introducing one electron close to the edge of a region of empty space, one would have electric fields turning into magnetic fields and vice versa ad infinitum. Possibly Maxwell's first reaction was that this was all so crazy his new expression just had to be wrong.

I like to picture Maxwell thinking at this point 'what does the mathematics say?' The mathematics of Faraday's law and his own, new version of Ampère's law were suggesting an apparently crazy situation when one switched from one equation to the other and back again. Why not see what happens, he thought, if the two equations are combined into one? The mathematical form of these two equations is complicated because the electric and magnetic fields vary in time and space. However, combining them together is a mathematical exercise which generations of physics undergraduates have worked through. In doing so, they have experienced some of the excitement of Maxwell's discovery.

When Maxwell first combined the two equations, he must have been intrigued to find that the single equation that results is a very particular type. It is what physicists call a *wave equation*. All waves, like waves on the sea or sound waves in the air, follow a similar equation. One important point about a wave equation, known to all physics students, is that the speed of the wave appears in the equation.

When Maxwell combined his new equation with Faraday's and found a single wave equation, I am certain the first thing he would have done was to search for the part of the equation that corresponded to the speed of the wave.

WHAT IS LIGHT?

It must have been with mounting excitement that Maxwell extracted the speed of the wave moving according to this wave equation. His change to Ampère's law was predicting a new type of wave. If experimentalists could find waves moving with this speed, his discovery would be confirmed.

When Maxwell extracted the speed of his new waves from the new equation, he must have had one of the greatest highs experienced by any physicist, theorist or experimentalist, in any age. The speed of this new type of wave was a very large number, which was very close to a number he knew very well indeed. It was an important number in a different branch of physics to electricity and magnetism.

It was the *speed of light*.

The waves that would confirm his idea had already been discovered. Their speed had been measured, quite accurately, just three years before. It couldn't be a coincidence that his waves moved with exactly the speed of light.

By adding one small, new contribution to a set of four equations, which physicists, quite rightly, now name after him, Maxwell had made three staggering discoveries, any one of which would be worthy of the Nobel Prize had he lived long enough.

Maxwell had explained how a time-honoured and apparently unrelated branch of physics works. *Optics*, the study of light, had suddenly become the study of *electromagnetic waves*. All optics from the largest telescope to the smallest microscope can, in principle, be explained with Maxwell's equations; though physics undergraduates will tell you it may not necessarily be the easiest way to solve a particular optics problem. Hereafter, physicists would consider sunlight an electromagnetic wave; a wave of time-varying electric fields turning into time-varying magnetic fields turning into time-varying electric fields ad infinitum.

Second, by uniting electricity and magnetism into a new force *electromagnetism,* Maxwell had reduced the number of fundamental forces then required to explain Nature. Physics had previously required three fundamental forces; gravity, electricity and magnetism. Now there were only two: gravity and *electromagnetism*.

Third, and unintentionally, he had unleashed a technology that was ultimately to give us, though only after two revolutions, which I will describe, most of our modern devices – from giant particle accelerators down to mobile phones and ultimately, I hope, to non-burning energy security.

Maxwell died in 1879, two years after the premiere of Tchaikovsky's *Swan Lake* and, sadly, eight years before electromagnetic waves were first deliberately generated and detected in a laboratory. He therefore could not have been aware of the extraordinary range of practical applications that were made possible by his discovery. Could he possibly have imagined that the dancing electric fields and magnetic fields are so perfectly choreographed by his equations that, a century after his death, his waves could faithfully transport the whole of *Swan Lake* into many twentieth-century living rooms with movement, colour and sound reproduced as well as in a concert hall? Equally unbelievably, by 2010 his waves would be conveyed along plastic fibres exactly according to his equations so that every word of *David Copperfield* could be downloaded almost instantaneously onto an e-book without a single full stop being displaced.

There is another reason why we need to understand the revolutionary nature of Maxwell's idea that light waves are more classical ballet than rock and roll. It is not simply because most popular physics texts gloss over this subject. We will then appreciate how shocking was the *next* change in physicists' picture of sunlight. In 1905 Einstein proposed a new picture of sunlight as revolutionary as Maxwell's had been 40 years earlier. At first sight, the pictures of these two scientific giants appeared incompatible, like punk rock meets *Strictly Come Dancing*. In one of the greatest paradoxes of physics, *both* pictures are correct.

The next unification

It is also important to appreciate the impact of Maxwell's unification of the two fundamental forces, electricity and magnetism,
into one force, electromagnetism. When physicists discover new
phenomena that cannot be explained in terms of existing fundamental forces, they invent new forces to explain them. As we
will see, in the 1930s the *strong* and *weak nuclear* forces were
discovered with traumatic effect. Following Maxwell's sensational
example, many generations of physicists have tried to *reduce*
the number of fundamental forces by unifying these new forces
with existing ones. It took just over a century of effort following
Maxwell's breakthrough before there was another unification of
fundamental forces.

In the late 1960s, electromagnetism and the weak nuclear force
were unified into the *electroweak* force – my old colleagues came
up with a descriptive but not particularly elegant name. In contrast
to Maxwell's breakthrough, this unification required the theoretical,
experimental and technological contributions from many hundreds
of physicists and engineers from many countries. As a young particle
physicist in the late 1970s, I was privileged to take a small part in
a big experimental team at CERN, which made a contribution to
the understanding of the unified force. It was a very exciting time,
but this unification was also very different from Maxwell's in that
it did not explain any other major areas of physics and it has
not, as yet, led directly to any practical, commercial application
of which I am aware. A chance encounter at that time, however,
did give me the opportunity to start a friendship with a Nobel
Prize-winner and also gave me an insight into the priorities of
the scientific press.

On a cold, grey November afternoon in 1979 I was giving an
undergraduate tutorial when our head of department burst into the
room with: 'Professor Abdus Salam has been awarded the Nobel

Prize for his theory of electroweak unification. There is a press conference in fifteen minutes. As you are the only experimentalist not in CERN today can you come along in case there are questions about the experimental confirmation?' I was extremely apprehensive but need not have worried. Abdus was a charming man, a great physicist and well able to handle all questions. We struck up a friendship as a result of this experience. Years later, Abdus was extremely supportive when I switched to solar cell research. He could see how important the technology could be for his native Pakistan.

During his presentation, Abdus recounted a delightful story about his daughter who was studying high-school physics. The teacher was talking about how stable the proton was. The daughter raised her hand. 'Please Miss, my father thinks that protons may decay after a very long time.' The teacher immediately responded with 'if you put down everything your father thinks you will fail your examination'.

I learnt a lot about the preoccupations of the scientific press when the time came for questions. I was expecting something along the lines of 'how does it feel to be the first person since Maxwell to unify two fundamental forces?' and 'how soon do you think there will be practical applications of your discovery?' How naïve I was. The first question was 'is Pakistan developing a nuclear bomb?' to which the answer was 'I have no idea'. The second question was 'did your daughter pass her exams?' to which his answer was 'Yes'. This was the part of Abdus's presentation that captured the headlines next day.

Elegant! But is it any use?

The first large-scale, commercial application of Maxwell's equations came in 1881, 16 years after his breakthrough. In that year the equation he had borrowed from Faraday was used to generate electricity for street lighting. It had taken exactly 50 years from when Faraday first proposed his law.

WHAT IS LIGHT?

In those 50 years engineers had found a practical way to get a time-varying magnetic field to pass through a loop of wire. This would produce an electric field in the wire which would pull electrons and make a current. An external source of power such as a steam engine rotated a circular loop of wire about its own diameter in the steady magnetic field of a permanent magnet.

Another practical development was to wind one single length of wire into a coil of many loops. If there are 100 loops in the coil, the voltage between the ends of the wire is 100 times larger than in a single loop. A device with a coil rotating in a magnetic field is called a *dynamo*. It converts the energy of rotation of the coil into the electrical energy of the electrons forced to flow in the wire by the electric field.

One breakthrough illustrates another example of the beautiful symmetries of physics. If the external source of rotation is disconnected and the ends of the wire connected instead to a battery so that a current is forced down the wire, *the coil rotates of its own accord*. In other words, the appropriate connections to an external electrical supply can turn a dynamo into an *electric motor*. The electrical energy of the electrons forced to move by the battery is converted into the energy of rotation of the coil and the motor can do useful work.

Neither the development of the dynamo nor the electric motor need have delayed the commercial application of Faraday's law. Faraday himself demonstrated a simple motor in 1831. The 50-year delay probably owed more to entrepreneurs and technologists thinking that the Industrial Revolution was proceding quite nicely thank you, fuelled by coal burning; who needs this new, clean but probably expensive electrical power?

It is likely that the Victorian fossil fuel industry saw electricity more as a threat than an opportunity. Support for this hypothesis comes from the observation that in 1881 the first commercial application of electricity was street lighting. The major cities of the UK all had their streets lit by gas produced from coal before Faraday's time. There is a parallel here with modern times. We will look later at the influence of the fossil fuel industry on the rate at which renewable technology is expanding today.

Though 50 years to the first exploitation of Faraday's law can hardly be described as a race, the front-runners were neck and neck in the final lap. A number of contestants were using water wheels in the English towns of Chesterfield, Godalming and Norwich to turn the coil in the dynamo and supply electricity to street lights. The main competitor, the American inventor Thomas Edison, used steam produced by coal burning to turn his dynamo. It powered street lights on Holborn Viaduct in London, only a mile or so (and 50 years) from Faraday's laboratory at the Royal Institution.

With great pleasure I can announce the winner of the race to provide the first public electricity supply. By just weeks in September 1881, it was renewably generated hydropower on the River Wey in Godalming!

Electricity generation has become more sophisticated over the last 130 years, but most electrical power today is still generated according to Faraday's law by rotating coils in a dynamo. What has changed is the nature of the technologies producing the rotation. The burning technologies – coal, gas, oil and nuclear – have dominated. A few countries like Norway, Switzerland and Italy have exploited their abundant hydropower resources. In recent years, the other solar approaches – wind power, wave power and tidal power – have increasingly provided the rotational energy of the dynamo coils. Not only do they not require fuel and generate far less greenhouse gases than all the burning technologies, they are often simpler technologies. This is because renewable sources usually turn the dynamo coils directly. There is no need to first heat water to produce steam, as happens with the burning technologies.

Solar cells are the first technology to make a major contribution to our electricity supply without rotating coils and without using Faraday's law. Solar cell technology is revolutionary though nothing actually revolves.

WHAT IS LIGHT?

From Maestro Maxwell to Marconi

The first time electromagnetic waves were generated and detected in a laboratory was in 1887. That was 22 years after Maxwell's discovery and, sadly, 8 years after his death. Heinrich Hertz was a German scientist with a very different background to Gauss as he came from a very wealthy home in Hamburg. His father was a lawyer and a senator.

Hertz was studying large electric fields in loops of wire. The fields were so large that when the circuit was broken a spark jumped across the gap. Hertz knew that, as the large electric field would then quickly fall to zero, the electric field would be changing fast with time. If Maxwell was right, this would produce a large time-varying magnetic field, which in turn would generate another time-varying electric field. Hertz hoped that the spark would produce electromagnetic waves, which he could detect.

How would Hertz prove that he had generated Maxwell's waves? He had an ingenious idea to find them. He arranged a second, very similar loop of wire some distance away, which also had a small gap. Just like a dynamo acting as an electric motor, Hertz assumed that if the first loop generated electromagnetic waves a second identical loop could detect them. He expected that the time-varying magnetic fields in the waves would produce a time-varying electric field in the second loop.

Hertz adjusted the size of the gap in the second loop until a small spark appeared. The loops of wire were the first examples of what were later to be called *wireless aerials*. The first loop was the *transmitter* and the second loop the *receiver* of electromagnetic waves.

Hertz showed that his new waves, which we now call *radio waves*, had very much longer wavelengths than sunlight. Apart from the longer wavelength, the radio waves were in all other respects electromagnetic waves like sunlight. The real clincher was that Hertz showed that his radio waves travelled with the speed of light.

Hertz's experiment is an example of a phenomenon known as *resonance*, which will be important many times in our story. Next time you tune your radio to the wavelength of your favourite station, reflect on the fact that you are repeating Hertz's experiment and proving Maxwell was right. What you are doing is tuning the electronic circuit in your radio until it resonates at the wavelength of the electromagnetic waves broadcast by the station. By adjusting the size of the gap, Hertz was ensuring resonance between his detecting loop and the transmitting circuit.

Amazingly, Hertz himself could see no practical application for his discovery. He considered it simply a confirmation of the ideas of the 'Maestro Maxwell'. It took a further 14 years effort in many countries before in 1901 the Italian, Guglielmo Marconi and his team transmitted the letter 'S' in Morse code from Cornwall to Newfoundland. Like Hertz, the Italian Marconi was also from a wealthy family. His mother was a Jameson; they owned the famous Irish whiskey distillery.

The fundamental problem of these early wireless experiments was how to pick up the information that the electromagnetic wave transmits to the second loop. The electromagnetic wave and the electric field in the second loop vibrate very fast. As a result, the electrons in the second loop move back and forth very fast too. No means of detection then available could measure such fast varying currents. As we will find, it took 100 years before a device was developed that could react fast enough to pick up the electron motion due to radio waves directly; the rest is mobile phone history.

It was another German physicist, Carl Braun, who came from the town of Fulda in the centre of Germany, who realised that what was needed was a device called a *rectifier*, which ensures electrons only move one way. The gap in Hertz's experiment achieved this, turning the back-and-forth motion of the electrons into a spark in one direction across the gap; but that is no way to run a radio station. Braun invented his own rectifier by making metal contacts to certain types of crystal. Other radio pioneers such as Marconi had their own rectifiers. All of them had their problems. In Braun's case, the performance of the metal to crystal contact was unpredictable, partly because no one knew how it worked.

The development of wireless received a major boost in 1904 when the English engineer John Ambrose Fleming, a professor at University College London and a consultant to Marconi, invented a very different and much more reliable rectifier. It was called a *vacuum diode*. It was made from a glass bulb from which the air has been evacuated. It resembled an old-fashioned filament light bulb, which pops when the glass shatters because there is a vacuum inside. The diode contains a heated wire (known as the cathode) like the filament in the old lamp bulbs. In the case of the diode, the heated wire gives off electrons, which are collected by a metal contact (known as the anode). These electrons form a current when an electric field is applied in one direction between the heated wire and the anode contact. No current flows when the electric field is connected in the opposite direction. Hence this device acts as a reliable rectifier.

An even bigger breakthrough followed two years later. The American engineer and entrepreneur Lee de Forest invented what is now called the *triode valve*. This was a vacuum diode with a third contact, which was a metallic grid inserted between the heated wire cathode and the metal anode. A second electric field between this grid and the cathode influences the flow of electrons on their way to the anode. A very weak signal (de Forest called them 'feeble' in his patent) from an aerial that was fed to the grid could then be multiplied many times in the current flowing between cathode and anode. This was the first example of an *amplifier* of the very small electromagnetic wave signals in radio transmissions.

With reliable rectification made possible by the diode valve and amplification by the triode valve, radio technology developed at a much faster rate and Braun's rectifiers were consigned to the science museums. They still had a niche market as the 'cat's whisker' in the small personal radios known as 'crystal sets', which older readers may remember.

The historical swings of the physics and technology pendulum are fascinating. In 1947 it was realised, at last, that Braun's rectifiers were made from impure semiconductor crystals. Pure versions of Braun's rectifiers then made a dramatic entry, and diodes and triodes were developed inside semiconductor crystals. Ironically, it was then the turn of the bulky, fragile glass diodes and triodes to end up in the science museums.

By the end of the nineteenth century, physicists were confident that the major discoveries of their times were already leading to a better life for humankind. Electricity generated with Faraday's idea was increasingly being used in home and industry. A start was being made in using Maxwell's electromagnetic waves for communication. But, even before the disaster of the First World War erupted, physics had encountered its own trauma, which would influence the way life developed in the second half of the twentieth century. Ironically, the trauma in physics was resolved by yet another revolution in the way physicists describe sunlight.

THREE

The Quantum Revolution

Those who are not shocked when they first come across quantum theory cannot possibly have understood it.

Niels Bohr

Cornish cliff-tops provide some of the most stunningly beautiful views in Britain. They also experience some of the wildest weather. Picture the 26-year-old Marconi, in a smart Italian fur coat, on the cliff-top at the turn of the century. His assistants, probably less well protected from the elements, are struggling to keep aloft hundreds of metres of wire, as rain and wind gust in from the Atlantic.

Imagine one of his assistants shouting, 'Sir ... I dreamt that some day ... thanks to your invention ... in 100 years' time ... a gentleman will have secreted about his person a small device ... to inform his wife when to have dinner on the table!'

Picture Marconi replying, 'Stop dreaming ... hang on to the wire ... I had a nightmare that someone made Herr Braun's rectifiers work with tiny aerials ... and history gave *him* the credit!'

Next time you use your mobile phone you might reflect on the fact that inside is an extremely compact version of the aerial Marconi and his team were struggling to assemble on the Cornish cliff-top. Your mobile phone, and the infrastructure that forwards your messages, both depend on much smaller and more complex versions of Braun's rectifiers. It was not until 1947 that it became clear that Braun's devices were made of a new type of electrical material: the *semiconductor*. I imagined Marconi's nightmare, but to some extent it came true. He had to share the 1909 Nobel Prize with Braun: one of the most far-sighted of the Nobel committee's decisions.

The science behind your mobile phone and laptop will be described in later chapters. It was developed during the semiconductor revolution in the second half of the twentieth century. In this chapter, I will explain the quantum revolution, which led to the semiconductor revolution. It was a seismic shift in physics ideas, which took place between 1900 and 1930. The ideas are not easy to understand and, as Bohr's quotation implies, many find them shocking. Yet they explain a simple equation, which could be crucial to preserving life on earth.

As I mentioned in the Introduction, the equation $E = hf$ is less well known to non-scientists than $E = mc^2$. It deserves more exposure as it is more relevant to our everyday experience. It started the quantum revolution, which led to the semiconductor revolution and hence to your laptop and mobile phone. It can ensure our civilisation is not destroyed by the explosions of bombs, reactors or waste resulting from $E = mc^2$.

First, what does $E = mc^2$ mean? The E represents *energy*. Energy comes in many different forms – electrical energy, solar energy, nuclear energy and energy of motion, like the energy of children playing party games. When nineteenth-century physicists added up all the forms of energy, including the energy of electrons in high voltage wires and water droplets in clouds (both are types of *potential* energy or energy of position) they made an extraordinary observation. The sum of all the different forms of energy in any isolated system, for example our universe, comes to a number that has remained the same for billions of years. This principle, the conservation of energy, is so important that nineteenth-century physicists gave it the grand title *the first law of thermodynamics*. The results of all experiments have been consistent with this law including studies made at the very highest energies at CERN.

The *m* in Einstein's famous equation represents *mass*. It is that property of an object that determines how heavy it is. To a physicist, heaviness or weight is the force of gravity on an object; the greater this force, the greater the weight of the object and the greater the mass of the object. Gravity is actually an extremely weak force, much weaker than the electromagnetic force described in the last chapter. However, the earth has so much mass that the force of gravity appears very strong with every object we drop, or every step we take.

In Einstein's equation $E = mc^2$ the *c* stands for the speed of light, which we met in Chapter 2. It is a very large number. This equation says that mass can be converted into enormous amounts of energy. The mathematics is clear; it says the energy stored in an object of mass represented by *m* can be found by multiplying *m* by this very large number *twice*.

Einstein didn't explain how mass could be converted to energy. As we will see, over thirty years later a few researchers would unexpectedly show how some of the mass of a nucleus could be converted into unbelieveable amounts of energy. Einstein was horrified when two bombs based on his equation destroyed two Japanese cities. But first, what prompted Einstein to develop $E = hf$, his other, less celebrated equation which can save us from destroying ourselves with $E = mc^2$?

The ultraviolet catastrophe

While Marconi and his team were struggling to master radio waves and the Cornish gales, European physicists were wrestling with theoretical problems concerning these same electromagnetic waves, but at the shorter wavelength of visible light. The physicists were trying to explain a phenomenon crucial to our story: why is sunlight golden? The physics of the nineteenth century was yet to give a satisfactory answer.

THE HISTORY OF THE SEMICONDUCTOR REVOLUTION

To understand the problem, we need to know a bit more about the waves discussed in Chapter 2. We met water waves on the sea, sound waves in the air and eventually electromagnetic waves. All these waves are a succession of peaks and troughs; even electromagnetic waves have peaks and troughs in the electric and magnetic fields. We described the wavelength – the distance between two successive peaks – now we need to consider the *frequency* – the number of peaks passing in a second.

Remember in the last chapter we saw how a vibrating violin string produces a sound wave. A violinist presses her finger down on a string so that, when bowed, the string vibrates and emits a note of a particular pitch. The listener says the note has a high or a low pitch. The higher the pitch, the higher the number of vibrations there are in a second and so the higher the frequency of the note.

Both wavelength and frequency of a vibrating violin string are determined by the distance between the bridge of the violin and the player's finger. The string cannot vibrate at those points. So the player's finger determines the wavelength of the vibration created when the string is bowed. As her finger moves closer to the bridge, the string vibrates with shorter wavelengths and the note has a higher pitch and frequency. The *shorter* the wavelength, the *higher* is the frequency of the note. Those glorious high notes are always played with the virtuoso's fingers closest to the bridge.

In the last chapter I described how the vibrating string produces sound waves. In fact, you would not hear the sound of the violin string on its own in the auditorium. The wooden body of the violin has been carefully crafted to amplify the vibrations of the string. This is another example of resonance. We discussed in the last chapter how important resonance was when Hertz detected the first radio waves. The body of the violin, and the air inside, vibrate with the same frequency as the bowed string. As a result, more sound waves flow out and the note sounds louder.

The notes from a wind instrument like a clarinet are produced rather differently. The molecules in the enclosed air vibrate with a wavelength and frequency fixed by which finger holes are open or closed. Some of these vibrations leak out of the clarinet and are transmitted through the air in the auditorium as sound waves. Again these sound waves set your eardrum vibrating with the same frequency as the note played on the clarinet.

Going back to the nineteenth century, physicists wanted to calculate the intensity of sunlight at each of the colours of the rainbow. The sun is golden because the electromagnetic waves are most intense at a wavelength which we call yellow. But why is sunlight more intense at yellow wavelengths rather than at red wavelengths or blue wavelengths? The radiant heater in a ceramic hob glows 'red-hot' because the electromagnetic waves it emits are most intense at red wavelengths. The surface of our sun is much hotter. Nineteenth-century physicists guessed that the colour of a hot object depends on its temperature.

It was clearly not practical to do experiments on the sun, so physicists set up artificial suns in the laboratory where they could control the temperature. Empty metal enclosures, called cavities, were constructed. They could be heated up and then held at known temperatures. The physicists knew that the hotter the cavity walls got, the more the atoms in the solid walls vibrate. Atoms in solids are held at fixed distances apart by the very strong forces that keep the solid together. However, the atoms can vibrate around these fixed positions. The hotter a solid gets, the more the atoms jiggle around. This is what physicists call *heat energy* – the energy of the jiggling motion of atoms about their fixed positions in a solid. On the other hand, the heat energy of the air is the energy of the random motion of the air molecules bouncing off each other more than a billion times a second as described in Chapter 2. The more the atoms in a solid jiggle around and the faster air molecules fly around, the higher the temperature.

The nineteenth-century physicists thought that the vibrations of the atoms in the cavity walls would excite electromagnetic waves inside these artificial suns. They didn't understand how this would happen. They incorrectly assumed it would be rather like the jiggling atoms in the solid walls bashing the gas molecules in the air inside the cavity. However, they correctly appreciated that, once the waves were excited, some of them would leak out of a small hole in the cavity. So the experimenters set up apparatus to measure the energy in the electromagnetic waves as they came out of the hole.

One of the most famous English physicists, the aristocrat Lord Rayleigh, calculated the energy of these electromagnetic waves. He was the next Cavendish Professor of Experimental Physics at Cambridge after Maxwell and the one before J.J. Thomson who discovered the electron. Rayleigh's answers were very different from the experimental results. The problem was acute. Rayleigh had a brilliant track record: he was the man who explained why the sky is blue. Where could he have gone wrong?

There were only two parts to Rayleigh's calculation. First, he pictured the electromagnetic waves bouncing back and forth within the cavity and occasionally leaking out through the small hole, like sound waves from a clarinet. Rayleigh knew how the wavelengths of sound waves in a clarinet are fixed by the distances between open and closed holes. He was, after all, author of the scientific best-seller *The Theory of Sound*. He fitted the electromagnetic waves into the cavity in a similar way. For example, *ultraviolet* (UV) is light we cannot see, though too much of it can harm our skin. UV light has much shorter wavelength than yellow sunlight. Therefore Rayleigh calculated that many more UV waves can be fitted into the cavity than yellow waves.

The second part of Rayleigh's calculation was to decide how much energy each wave must carry. Since they didn't know how the jiggling atoms in the cavity walls excited the waves, nineteenth-century physicists like Rayleigh could only guess how much energy each wave carried. What they *did* know was how solid material like the cavity walls behaved when heated. Physicists can explain how much hotter a cavity gets as the atoms jiggle around more. They do so by assuming that, on average, all the atoms in the material jiggle around with the *same* energy of motion. So the assumption Rayleigh made, backed up by many other physicists at the time, was that each wave in the cavity had the same energy on average, just like the average energy of motion of an atom in the wall.

Here comes the big problem. If all waves had the same energy, then the more waves there were in the cavity at one wavelength the more energy the experiments should measure emerging from the hole at that wavelength. Rayleigh had calculated that there were more, short wavelength UV waves in the cavity than yellow waves. Hence he was predicting that the experimenters would measure much more energy emerging in the UV than in yellow waves. This was the exact opposite of what was observed. His calculations predicted the sun should emit so much UV that the evolution of life on earth would have very been difficult. This major difference between Rayleigh's theory and the measurements was so serious that it was known at the time as the *ultraviolet catastrophe*.

Finding a solution to this problem produced an equation that started a scientific revolution and may yet save our civilisation.

Catastrophic problems lead to revolutionary solutions

At the turn of the century, a German physicist, Max Planck, suggested where Rayleigh might have gone wrong. Planck was the son of a professor of constitutional law at the University of Kiel. He decided that an authority on sound waves like Rayleigh could hardly have gone wrong in counting waves in the cavity. The error must have been in his second assumption – that all waves had the same energy on average as the atoms in the cavity walls.

THE HISTORY OF THE SEMICONDUCTOR REVOLUTION

Planck decided to look in more detail at how the energy of motion is shared among the atoms jiggling in the cavity walls and also in the air molecules flying around inside cavities. In both cases it is true that many molecules have energy of motion around, or just below, the average energy. However, there were some with above average energy and a few with very high energies. Gas molecules divide up the available energy like the population shares a nation's wealth. Most people have around or just below the average income. There are not many very rich people, but some are so wealthy they make a difference when calculating the average income.

Planck decided that Rayleigh should have used this more detailed picture to fix the way the energy was shared by waves in the cavity. Planck's next inspired step was to propose that UV waves have much higher energy than yellow waves. Then, according to this picture, there would be many less UV waves than yellow waves in the cavity; just like the small numbers of very rich people.

How did Planck decide that these dangerous high frequency UV waves carried higher energy than yellow waves? His final stroke of genius was to think up a simple equation, which showed that the higher the frequency of an electromagnetic wave the higher is its energy. Appropriately published in 1900, as the old century ended, his choice was an equation which was to spark a revolution, sweep away much of nineteenth-century physics and provide the key to sustainable electricity generation:

$$E = hf$$

E represents the energy of an electromagnetic wave. The equation says that to calculate this energy you take the frequency (f) of the wave and multiply it by a fixed and very small number represented by the symbol h. The higher the frequency, the higher is the energy of the wave. UV light has at least twice the frequency of yellow light, so UV light has at least twice the energy of yellow light.

Planck put his new equation into the detailed formula for the way gas molecules share energy. When he did this, he found his calculations agreed with experiments, well within the experimental errors and far better than Rayleigh's calculations. Planck correctly predicted that the sun radiates much more energy in yellow light than in the UV. The German had averted the ultraviolet catastrophe.

Planck only got this excellent agreement with experiment if the number h was exceedingly small. To write the number represented by the h without using scientific shorthand you need 33 zeros after the decimal point. Here it is: $h = 0.00000000000000000000000000000000006626$. Later we will see the part played by this minuscule number in an amazing coincidence. The coincidence links the most important property of the silicon crystals in your computer, the fact that our sun is golden and our sustainable future.

Planck had explained why the sun is golden, how much energy it emits at each wavelength and averted the ultraviolet catastrophe. But what did this equation mean?

Planck himself, and many other physicists at the time, thought $E = hf$ was a mathematical trick which ensured that the detailed formula for energy sharing gave results that agreed with experiments. Some thought h was so small it probably didn't matter. But as the quantum revolution developed, this very small number kept turning up, often in mysterious circumstances. It is now called *Planck's constant*. It is one of the most important numbers in physics, on a par with the speed of light. But the equation $E = hf$ is not usually named after Planck alone. It took Einstein's genius to explain the revolutionary nature of this equation.

Meanwhile, in a patent office in Berne ...

Was $E = hf$ just a mathematical trick? What was needed was a physical picture of what the equation meant. As so often in physics history, Albert Einstein provided the physical picture.

In 1905 Einstein was not a practising academic physicist. He was a young patent officer in Berne, Switzerland. But in that one year, in his spare moments, he found time to publish a number of groundbreaking papers. His interpretation of $E = hf$ was but one. He also published his first papers on his theory of relativity and on $E = mc^2$. Clearly patent work must have been quiet in Berne in 1905.

Einstein's interpretation of $E = hf$ started the quantum revolution. As so often with revolutionary new ideas, they appear obvious in retrospect. Einstein reasoned that Planck had shown that he could explain why the sun is golden if the waves in a cavity share energy like the molecules or the particles that make up a gas. Perhaps nature was trying to tell us something. Perhaps electromagnetic waves *are* particles: tiny pieces or packets of energy. If so, the equation $E = hf$ was not a mathematical trick but an absolutely crucial definition of how big these packets are. Einstein proposed that the size of a packet of energy E when an electromagnetic wave has frequency f is given by $E = hf$.

Einstein named a tiny packet of wave energy a *quantum* (or *quanta* in the plural). The fact that the sun appears golden now has a simple explanation. The sun emits more quanta that our eyes decide are yellow than any other quanta.

Why had no other physicist come up with this idea in the five years since Planck had first published this equation? Clearly Einstein was a genius in deciding that electromagnetic waves had to have particle properties, but there also is some history here. Remember in Chapter 2 how Young had resolved the physics controversy, which had raged for over a century, about whether sunlight was made up of particles or waves? The particle picture of sunlight had been championed by none other than Sir Isaac Newton, the grand old man of English physics; he had called the particles *corpuscles*. Young's experiment, and Maxwell's elegant explanation of sunlight as an electromagnetic wave, had appeared to settle the controversy; sunlight consisted of waves. Now this young, upstart patent officer was proposing that the biggest physics controversy for more than two centuries should be settled as a tie! An electromagnetic wave has *both* wave *and* particle properties.

Many physicists found this difficult to accept. Particles like protons and electrons, which they pictured as tiny snooker balls, seem so different to electromagnetic waves. But Einstein convinced doubters, and the Nobel Prize committee, by explaining some puzzling experimental results.

The odd results were from experiments started towards the end of the nineteenth century by the English physicist J.J. Thomson, the discoverer of the electron, whom we met in the last chapter. Thomson studied the electric current flowing between the metal contacts – the cathode and anode – in a vacuum tube. He showed that the current consisted of the flow of electrons. The diode and triode valves were based on his apparatus. Fleming and de Forest used heated cathodes to get more electrons out of the metal. Other experimenters shone light on the cathode to try to get more electrons. They found that, strangely, when blue light shines on the cathode more electrons emerge but when a red light shines there are no electrons to be found, no matter how bright the red light is.

Einstein's idea explained this unexpected result brilliantly – in more ways than one. He assumed that a single quantum of red light has too little energy to give one electron in the cathode enough energy to get out of the metal. According to $E = hf$, a blue quantum of light will have twice the energy of a red quantum. Using the equation, Einstein correctly predicted how many electrons are emitted in a bright blue light if one blue quantum has enough energy to push one electron out of the contact.

It was his ideas about $E = hf$, rather than his papers on relativity, that won Einstein the Nobel Prize in 1921. The Nobel committee showed great foresight in deciding that $E = hf$ would have more beneficial applications than $E = mc^2$. That was despite Einstein's more famous equation promising more powerful explosions than the dynamite with which Alfred Nobel made his money.

Einstein's idea that electromagnetic waves also behave like particles marked the starting point for quantum theory, which proved to be a revolutionary development in physics. This idea is now fully accepted by physicists. So much so that the name *quantum* for a packet of electromagnetic energy has been replaced by a particle name; it is called a *photon*.

When first meeting the idea, most people are shocked that sunlight can be described both like a collection of snooker balls and also like waves on water; this must be the ultimate mixed metaphor! But we *are* talking about the ultimate; we are talking about the fundamental particles that make up sunlight. Snooker balls and water waves behave as they do because they are made up of trillions and trillions of atoms interacting according to the laws of physics. When physicists get down to studying at the scale of an atom, why should *one* atom or *one* photon behave like an object made up of trillions and trillions of atoms? Quantum physics, and modern communications, would have been much less interesting had it turned out sunlight behaves simply like a lot of snooker balls or simply like a lot of water waves.

The infrared catastrophe

The next success for the equation $E = hf$, came just before Europe was engulfed in the First World War. Niels Bohr, a Danish theoretical physicist, like Max Planck came from an academic family. This would become increasingly the case among the twentieth century's scientific trendsetters. Bohr was trying to work out if Einstein's quantum idea and the equation $E = hf$ could avert a second potential catastrophe in physics. The way he resolved the problem placed him at the head of what by now was becoming a quantum revolution.

Bohr was interested in how light is emitted by individual atoms rather than from the surface of the sun. Atoms glow with a variety of interesting colours when they are heated. This can be seen in the orange sodium street lighting and the many colours of neon signs. The sodium and neon refer to different gaseous elements inside the lamps. Remember that an element is a particular atom with a particular number of electrons.

If the one type of element is heated and the wavelengths of the light emitted are measured, the light is found to be built up of a distinctive pattern of colours and frequencies. This pattern is particular to each element. Just as a chord played on a piano is built up of notes of definite pitch, the light emitted by a particular element consists of a unique pattern of frequencies. Think of the spectrum of heated sodium as the chord of D major and the neon spectrum as the chord of G7 say.

Bohr wisely decided to study hydrogen, the simplest atom, first. Its nucleus is just one proton and it also has one electron. He guessed that the pattern of the light emitted by hydrogen, the distinctive chord of hydrogen, would provide clues to the structure of the atom. The picture of an atom had recently changed dramatically as a result of experiments led by Ernest Rutherford, a New Zealander. Rutherford's mother was English and a teacher. His father was a Scottish wheelwright. Both had emigrated to New Zealand and Ernest was the fourth child of twelve. Rutherford's definitive experiments were performed at the University in Manchester before he followed in the footsteps of Maxwell, Rayleigh and Thomson by becoming Cavendish Professor at Cambridge.

Rutherford's experiments showed that the atom was not, after all, like an extremely small snooker ball as generations of physicists had thought. He showed that nearly all the mass of the atom is concentrated in an even smaller nucleus right at its heart.

Bohr had worked for a time with Rutherford in Manchester. They both thought that the atom might resemble a tiny solar system. The very small heavy nucleus would take the place of the sun, with the much lighter electrons orbiting the nucleus like planets. This picture of the atom had a problem, which threatened to become a second catastrophe for physics.

As we saw in the last chapter, the electron has an electric charge; a planet does not. Maxwell said that a charged particle, like an electron, should radiate electromagnetic waves when it moves in an orbit. So far so good; perhaps that is how atoms emit light. However, as the electron radiates, it loses energy. If an atom is really a miniature solar system, then Bohr would expect the electron to behave like a planet. If a planet loses energy, it moves closer to the sun. Hence, as the electron radiates away its energy, it should get closer to the nucleus. Calculations showed that an orbiting electron would spiral down until it hits the nucleus in much less than a second.

So the picture of the atom as a miniature solar system had a major problem. If it were correct, every atom created in the universe should have vanished within a fraction of a second. Trillions and trillions of puffs of light with longer wavelength than red would have been emitted. Hence the problem was called the *infrared catastrophe*. This was as serious a problem for the new quantum ideas of the early twentieth century as the ultraviolet catastrophe had been for nineteenth-century physics.

Back in Copenhagen after his visit to Manchester, Bohr dealt with the potential catastrophe brilliantly. The atoms which make up our planet have existed for billions of years so some revolutionary ideas were necessary. No, said Bohr, electrons clearly do not lose energy, slow down and crash into the nucleus of the atom. Their orbits must be fixed in some way. Bohr proposed that electrons are only able to orbit at certain distances from their nucleus. As long as they remain in these orbits they cannot radiate electromagnetic waves.

But heated atoms clearly do emit electromagnetic waves. How come? The creativity of Bohr's thinking is shown in his answer. An electron must jump between orbits. The only way an electron could lose energy would be if it jumped from a higher energy orbit to a lower energy orbit closer to the nucleus. The energy lost in the jump would be emitted as a packet of energy. Perhaps, Bohr thought, this energy is radiated as a quantum of energy, $E = hf$, as Einstein had proposed?

Bohr's electron orbits are fixed. But how does nature decide the distance at which an electron orbits a nucleus? To explain this, Bohr came up with two more original ideas; not that he realised how revolutionary these ideas were at the time. He decided that each orbit should be labelled by a whole number. He labelled the orbit closest to the nucleus with the number 1, the next nearest orbit with the number 2, the next with the number 3 and so on. These labels were the first examples of what are now called *quantum numbers*. This may appear a simple idea but it had revolutionary consequences. Bohr used these whole numbers in a new equation relating two important quantities: the energy of motion of the electron in the orbit and the distance from the nucleus to the orbit. In an inspired moment, he also included Planck's very small and still mysterious number h in his equation. He then put this equation and Einstein's $E = hf$ into Newton's 200-year-old formula describing how planets move.

The result of this mathematical exercise must have given Bohr an experience every bit as sublime as Maxwell's. Bohr's calculations were in very good agreement with some extremely precise experimental results, which had already been published. They agreed with the pattern of light from the hydrogen atom – the notes of the hydrogen chord. Bohr's picture of electrons jumping between fixed orbits not only had to be correct, it had averted the infrared catastrophe.

Sadly, the development of these mind-blowing ideas was held up by the First World War – 'the war to end all wars'. There was no input from the new physics to the war effort and only marginal input from new technologies such as the internal combustion engine, aircraft and wireless. These had all been developed in the first decade of the century. Chemistry had horrific input with the use of poison gas on the battlefield. The war was fought to a stalemate in the trenches with heroic contributions from foot soldiers using mass-produced versions of the traditional weapons of war: guns, grenades, shells and machine guns.

Take-off for the quantum revolution

The First World War delayed the quantum revolution but did not stop it. In 1922, Louis de Broglie, a member of the French aristocracy, came up with an idea in his PhD thesis, which reignited the quantum flame. Maxwell had been the first to show the importance of symmetry in physics. Einstein had shown that electromagnetic waves behave like particles. The question de Broglie addressed in his thesis was: is physics symmetrical so that *particles like the electron behave like waves*?

It sounded crazy; but then so had Einstein's idea at first. De Broglie tried to picture an electron in an orbit round the nucleus behaving like a wave in a violin string. Electron wavelengths would have to fit into the circumference of the circular orbit like waves fit between the finger and bridge of a violin. De Broglie realised that fitting whole numbers of waves into the orbit would mean the circumference of the orbit could only take on certain values. Perhaps this is why electron orbits are only found at certain distances from the nucleus? Bohr's idea had never been satisfactorily explained.

To check this, de Broglie needed to find a formula for the wavelength of an electron when it had a particular energy. He assumed that the higher the energy of the electron the shorter the electron wavelength. This was consistent with Einstein's idea that the higher the photon energy, the shorter the photon wavelength. He also made sure his formula for the electron wavelength contained Planck's tiny number, which is symbolised by h.

When de Broglie used his new formula to fit electron waves

round the circumference of a circular orbit, he must have had his own Maxwell moment. He found exactly the same distances from the electron orbits to the nucleus as Bohr. De Broglie's idea that the electron was a wave explained why the electron orbits were at certain fixed distances. He had also explained why the quantum labels go up by one whole number with each orbit. This was because one more whole wave could be fitted into the orbit circumference each time. Here was proof that electrons were both particles and waves.

De Broglie's idea was a fundamental change in our view of matter. One of the most revolutionary ideas in physics originated from two aristocrats. The 7th Duke de Broglie fitted electron waves round an orbit just like the 3rd Baron Rayleigh fitted electromagnetic waves into a cavity.

More revolutionary physics ideas followed in quick succession in the mid-1920s. The first to build on de Broglie's idea was an Austrian, Erwin Schrödinger, who probably inherited his scientific interest from both parents. Schrödinger's father married the daughter of his chemistry professor. Schrödinger used de Broglie's idea to produce a formula to explain how electrons behave in atoms and in solid material. The formula is *Schrödinger's wave equation*. Naturally it includes *h*, three times in fact.

When his formula was applied to the hydrogen atom, Schrödinger predicted electron behaviour more bizarre than Bohr or de Broglie's wildest dreams. Rather than a wave at the circumference of the orbit serenely encircling the nucleus, the electron, according to Schrödinger, is a pulsating, fuzzy ball filling most of the space between the orbit and the nucleus. Even more mind-blowing, the mathematics says that in the second orbit the electron can take the shape of a hollow pulsating, fuzzy shell with another concentric fuzzy ball inside. Not only that, but a pulsating, fuzzy dumb-bell and two types of fuzzy doughnut are possible; a total of four different electron shapes for the second orbit. In the third orbit, there are four even more bizarre pulsating, fuzzy shapes. Shocking this may be, but it works. These shapes explain much of the detail of how atoms bond to form molecules.

Your mind is spinning? So are these electrons! In 1928 the English theoretical physicist Paul Dirac came up with an equation more consistent with Einstein's ideas about relativity than Schrödinger's formula. Dirac's mother was English and his father was Swiss. He went to school and university in Bristol before completing a PhD in Cambridge. Dirac's equation suggests that electrons are not just pulsating, fuzzy balls, dumb-bells and doughnuts, they are also spinning like tops.

Dirac's idea is not easy to reconcile with some of Schrödinger's pictures of the electron, but is simpler to appreciate with Bohr's view of the electron as a tiny planet. Bohr's picture still gives the right results in some cases. The earth spins in its orbit round the sun, turning night into day. The earth, however, only spins one way. Electrons, said Dirac, can spin both ways: east–west and west–east. This was a revolutionary proposal, in more ways than one.

Even more bizarre electron behaviour was predicted by the German, Werner Heisenberg in his famous *uncertainty principle*. Many will have heard of Heisenberg from Michael Frayn's play *Copenhagen*. The plot exploits the uncertainty about what was discussed when Heisenberg came to meet Bohr in Nazi-occupied Copenhagen in 1941.

The uncertainty in quantum physics, which Heisenberg quantified, results from an electron having both wave and particle natures. This makes an electron particularly elusive. The *more certain* you are *where* an electron is, the *more uncertain* you are *what energy* of motion it has. It appears crazy, but Heisenberg says that the more you confine an electron the more it jigs around and the higher its energy of motion. It is not easy to find an everyday analogy for this quantum behaviour, but in some respects an electron resembles a jelly. Try to squeeze it, to be certain where the jelly is, then it slithers all over you with more energy of motion. Pour it into a tray so large you are less certain where the jelly is, and it sets with less energy of motion.

Heisenberg represented this latest odd electron behaviour in a formula, which naturally includes Planck's very small number h. This was rapidly becoming less mysterious; by now every self-respecting quantum equation just had to include it.

By the late 1920s, experimenters were able to clearly demonstrate that electrons behave like waves. They could pass a beam of electrons through a crystal and observe behaviour like a beam of electromagnetic waves. One of these experimenters, the Englishman G.P. Thomson, later head of physics at my own university, Imperial College London, was the son of J.J. Thomson, who had discovered the electron. Hence the father showed that the electron was a particle and the son showed that the electron was a wave. Some hold that this was evidence of a Thomson family feud. I see it as the father and son demonstrating very different aspects of the electron reality.

But is it any use?

The two main quantum ideas were truly revolutionary within the realm of physics: electromagnetic waves behave like particles and the particles that make up matter also behave like waves. These ideas describe the exotic behaviour of electrons inside atoms, but do they have any practical application? In the next chapter, I will describe how quantum ideas led to one devastating application: nuclear weapons. But the first practical use of quantum ideas was in explaining much of chemistry, in particular how elements (atoms with a specific number of electrons) form molecules. This *quantum bonding* is rarely explained in popular science books about quantum theory, but it is crucial to the workings of the semiconductor crystals in solar cells and in your mobile phone.

Could the electrons in the orbit furthest from the nucleus be responsible for how atoms bond to form molecules? Chemists already knew that there is a family of elements that do not bond with *any other elements at all*, not even other atoms in their own family. The family is called the *noble gas* family, because it stays aloof from bonding with other atoms. The chemists decided there must be something special about the electrons in the outer orbit of these atoms, since they could never be persuaded to bond. It appeared to be because the outer orbit is full of electrons.

The first noble gas family member is called helium. It has two electrons. Chemists assumed they are both in the first electron orbit. They decided that two electrons in this inner orbit make for a full orbit and a stable atom that does not react with other atoms. The next noble gas family member is neon with ten electrons. The chemists had guessed correctly that there was only room for two electrons in the first orbit. Hence eight of the ten electrons must go into a second orbit. If there was only room for eight electrons in the second orbit, then neon would, like helium, have a full outer electron orbit. This might explain why neon is also stable and non-reactive, just like helium. Argon, the next noble gas family member, has eighteen electrons; that is eight more than neon. If these extra eight electrons fill up a third orbit, then helium, neon and argon all have full outer electron orbits. Chemists decided there is something stable and non-reactive about outer electron orbits containing 2, 8 and 8 electrons.

This conclusion was supported by the observation that the family of elements with one more electron than the noble gas family behave very differently to the noble gases. All family members with one electron more than a noble gas are highly active and bond with many other elements. Having only one electron in an outer orbit makes an element extremely reactive.

Chemists found that, thanks to these quantum ideas, they could now explain the formation of all sorts of molecules. Elements with outer electron orbits containing *fewer than* 2, 8 or 8 electrons rush around trying to find other atoms with unfilled outer electron orbits. The atoms then share, sacrifice or steal electrons from and to each other, so that a complete electron orbit containing 2, 8 or 8 electrons is formed. A full electron orbit with 2 or 8 outer electrons orbiting all the atoms is so stable it can hold all the atoms together as a molecule.

As an example take the most abundant molecule on earth: water. It is the key to life and to hydroelectricity, one of the solar technologies which can help save life on earth. There is one chemical formula everyone knows: water is H_2O. This formula means every water molecule contains one oxygen atom and two hydrogen atoms. Quantum bonding explains why. Oxygen has eight electrons. Two of these are in the first electron orbit, like helium, and six in the second. An oxygen atom needs two more electrons to have eight electrons in the second orbit to end up like neon. Any two hydrogen atoms that pass close by are only too happy to share their two electrons and settle in a threesome with the oxygen atom. Hence the three atoms are bound together by a full and stable outer orbit of eight electrons.

The 2, 8, 8 mystery

The new quantum ideas explained much of chemistry, but physicists were more intrigued by two questions. Why are orbits with 2, 8 and 8 electrons full? Why does having full orbits ensure the atoms and molecules are stable?

Physicists got really worried when chemists insisted there were only two electrons in the first orbit. This posed a problem that reminded the physicists of the infrared catastrophe. The electron has its lowest energy in the first orbit. Faced with a choice between high energy and low energy situations, nature (unlike many politicians) prefers the lower energy one. This is like a ball placed carefully on a table so it is stationary. Blow on the ball or nudge the table and nature will soon demonstrate it prefers the low energy and more stable situation of the ball nestling in a corner on the floor with lower energy of position. On approaching a nucleus, every electron prefers to jump down through the orbits until it reaches the first and lowest energy orbit. So all the electrons should be in the first orbit; not just two.

It was the Austrian physicist Wolfgang Pauli who took the first step in solving this problem. He, like Schrödinger, was born in Vienna and he was one of many quantum physicists who benefitted from working with Bohr in Copenhagen. As early as 1925 he explained part of the 2, 8, 8 mystery using quantum numbers as Bohr had done more than a decade earlier. A year later his fellow Austrian Schrödinger came up with his fuzzy electron pictures but the message remained the same. Schrödinger's wave equation showed that an electron takes on one shape in the first orbit (fuzzy ball) and four shapes for the second and third orbits (fuzzy ball and shell, dumb-bell and two doughnuts). That is a sequence of 1, 4 and 4 unique shapes. Multiply by two and you get the mysterious sequence 2, 8, 8.

Pauli's *exclusion principle* can be understood by assuming that electrons are all equal when roaming freely, but they are only allowed in an atom if they take on a shape that is different to any electron already in the atom. If all four shapes are already taken in the second orbit say, a new electron can only lower its energy by entering the atom in the empty third orbit. This is like the ball resting on the floor unable to lower its energy of position any further.

Pauli's principle suggests the numbers should be 1, 4, 4 rather than 2, 8, 8. His idea was only fully accepted when in 1928 Dirac came up with his theory that electrons can spin in two directions. If two electrons were vying for a place in an orbit that required they take one particular shape, they would *both* be allowed in if they were prepared to spin in opposite directions. Pauli's exclusion principle was then extended: electrons are only allowed into an atom if they take on a unique shape, *or spin in a different direction*, to electrons already in the atom. The mystery of the sequence 2, 8, 8, so important to chemists, had been explained.

Pauli's exclusion principle is vital to explaining how the

semiconductor crystals in your mobile phone and laptop work. There is an irony here. Pauli is reported as saying in 1931 that 'one shouldn't work on semiconductors, that is a filthy mess; who knows if they really exist '.

To help to understand Pauli's principle, here is an analogy prompted by my grandsons. (By chance, two of them attended Dirac's primary school in Bristol.) Imagine you are organising a game of musical chairs at a children's party. You are at the stage where you are left with a ring of eight chairs facing outwards. You would rather not see your best dining chairs ruined, so you have cobbled together an odd assortment of chairs; four pairs in fact, each pair with a different shape. The chairs in each pair are a different colour to represent the ways electrons can spin. The children represent the electrons. They circle the ring of chairs while the music plays. When the music stops, there is a mad scramble for the chairs. Once on a chair, the children will take the shape dictated by the chair. The Pauli principle insists that any child can sit on any chair (all children and all electrons are equal) but *no sitting on laps*! The end result of the mayhem is eight, contented, seated children, reluctant to get up when the music restarts. There are a few disappointed and hyperactive children, excluded from the game, like the electrons in the next, empty orbit.

Why are full electron orbits so stable?

There was still one more question for the physicists. How come electron orbits with two or eight outer electrons are so stable they can bond atoms together? One of my heroes from my particle physics days, the late Richard Feynman, provided the most stimulating mathematical account of quantum bonding using the simplest example: two hydrogen atoms forming a hydrogen molecule.

When two hydrogen atoms share two electrons the total energy is lower than when the two atoms are separate, each with one electron. As noted earlier, when faced by two options, a high energy one and a low energy one, nature chooses the lower energy option. But why does the electron sharing option have lower energy? Feynman explains that when two hydrogen atoms get very close together the electrons on both atoms can jump on to the other atom and then back again. Remember the electron is both wave and particle. Because the electrons are particles and there is only one electron on each atom, there is room for a second electron as long as both electrons are spinning in opposite directions.

But why should the electrons bother to jump? Now the wave-nature of the electron matters. Both hydrogen atoms are identical so the wavelength of an electron in the same orbit on either atom will be identical. Just like Hertz detecting his radio waves, this is the perfect situation for the electron on one atom to set up a resonant wave on the other. This electron wave will then resonate back again onto the original atom.

But that still doesn't explain why electron sharing is the lower energy option. Feynman explains this with Heisenberg's uncertainty principle. When the two electrons are spread over both atoms, the positions of the electrons are *less certain* than when they were each on their own atoms. You are more certain where each electron is when they are on their own atoms. Remember that an electron is like jelly. When you squeeze it into a fuzzy ball, it jiggles around. It then has higher energy of motion than when it spreads itself over both atoms. Electron sharing is the preferred lower energy situation.

The musical chairs analogy can also help explain how two hydrogen atoms bond. Assume the game is nearly over and there is one pair of chairs left. Anyone who has organised this game knows that in the last round the two chairs must be well separated to keep the remaining children circulating while the music plays. Each chair now represents the fuzzy ball shape in the first electron orbit of the hydrogen atom. The two chairs have the same shape but different colour representing the different ways of spinning.

The music stops and two lucky children sit down, one on each chair. They are probably jiggling around, keen to start the last round. But what happens if the two children decide they would prefer not to compete in the final round with one chair but instead share the prize? Perhaps they refuse to get up when the music starts. With the two chairs separated you could reason with them – divide and rule! You could even try tipping them off the chairs in a friendly fashion. But, if they shuffle the chairs together and lay across both chairs, you are in trouble. You cannot pull the chairs apart without risking some harm. The two atoms are bound together.

Doughnuts, jelly, spinning tops and musical chairs; quantum theory is child's play!

Quantum bonding demonstrated by a quantum revolutionary?

While a young postdoc in Berkeley I missed out on an opportunity to meet one of the greats of quantum physics. One of the highlights of my two years there was going to be presenting some results at a conference in Caltech, where I hoped to get the chance to hear one of my scientific heroes, Richard Feynman. In fact, the conference turned out to be one of the most traumatic in my research career. I was presenting work trying to reconcile apparently inconsistent data from different experimental teams with a quantum model. This model had been developed by my then boss and subsequent, long-term friend: the late Gerson Goldhaber. There was controversy about whether the data showed that one sub-nuclear particle was actually split into two different particles. On the morning of my presentation, I learned that the talk immediately after mine was from a team with some new experimental results. These cast strong doubts on the data from one of the major laboratories in my comparison. This was a formative experience for a young researcher.

THE HISTORY OF THE SEMICONDUCTOR REVOLUTION

I was still preoccupied later in the conference when, wandering through an emptying lecture theatre, I passed a group huddled around a sun-burnt, male physicist with an impressive head of hair lounging across a number of seats. The attendees were hanging on the speaker's every word. I assumed the speaker was a type common in Berkeley at that time: an over-age graduate student, who had still not submitted his thesis after 10 years or so. He would then have to leave the idyll that was Berkeley and run the risk of being drafted to Vietnam. The speaker was giving the impression to his listeners that he knew it all. I was some way past the group before it struck me ... I had seen a photo of that face ... he *did* know it all! I was too embarrassed to retrace my steps to join the throng to find what he was talking about. I like to think, however, that by reclining over a number of seats in the auditorium, Richard Feynman was illustrating the subtleties of quantum bonding.

Quantum bonding and fossil fuels

Quantum bonding explains the cornucopia of carbon-based molecules that make up our plastics, our fossil fuels and the even more complex molecules of life. The way quantum bonding explains all these amazingly different molecules was first worked out in the 1930s by another of my heroes, the American Linus Pauling. He is the first person to have gained two unshared Nobel Prizes, for Chemistry in 1954 and Peace in 1962. Not only was Pauling the first to explain much of chemistry in terms of quantum bonding, he was also involved politically. His pioneering efforts to alert politicians to the radiation dangers from atmospheric testing of nuclear weapons eventually led to international agreements that banned such tests. Pauling's life, his work and ideas, exemplified the intertwined nature of science and politics, as must any history of solar energy.

The amazing diversity of carbon molecules is the result of every carbon atom having four electrons in its outer orbit but preferring to have eight. There are many ways this can be achieved with other carbon atoms. For example, two of the electrons can be shared with two hydrogen atoms and the remaining two with two neighbouring carbon atoms. The original carbon atom then thinks it has an outer orbit of eight electrons. Also, both hydrogen atoms think they have a stable orbit of two electrons like helium. But what do the two neighbouring carbon atoms think? They have shared one of their original four electron atoms with our original carbon atom; so they still each have three electrons to share. They too can share one electron with another carbon neighbour. Furthermore, they can share two electrons with another two hydrogen atoms.

In this way, a long chain-like molecule with a backbone of carbon atoms is built up. The hydrogen atoms hang off the backbone. The whole molecule is bonded together because every carbon atom thinks it is surrounded by a full orbit of eight electrons and every hydrogen atom thinks it has a full orbit of two atoms. These long chains are called *hydrocarbon* molecules. They are the basis of our fossil fuels – coal, oil, gas – and the polymers and plastics derived from them. Even the molecules of life itself are based on carbon. Proteins and DNA are highly complex carbon chains, with some of the hydrogen atoms replaced by other elements keen to share electrons.

Sunlight formed all these chains by photosynthesis over two-and-a-half billion years. This was how the vegetation was formed which decayed to produce the hydrocarbon molecules in the coal and oil we are burning as if they were inexhaustible. In a car engine the bonds in these long chains are broken in the presence of oxygen from the air. This releases simpler, stable molecules like carbon dioxide and water and excess energy in the form of the energy of motion of these molecules. The fast-moving molecules push the pistons and drive the car down the motorway. In a car engine this bond-breaking is controlled. The uncontrolled burning of hydrocarbons can lead to explosions as in the BP Deepwater Horizon disaster in the Gulf of Mexico in April 2010.

The devil's bargain

By the early 1930s the basic ideas of the quantum revolution were in place, and physicists and chemists were applying these ideas in many areas. In 1931 the English quantum physicist Alan Wilson, working in Germany with Heisenberg of uncertainty fame, published an explanation of how a semiconductor crystal might work. Yet the semiconductor revolution did not start for another 16 years. In the next chapter, we will see how two cataclysmic events, which occurred close together in late 1932 and early 1933, delayed the start of the semiconductor revolution. We will also find how nuclear physics and the equation $E = mc^2$ led to a devastating practical application of quantum ideas.

FOUR

Brighter than a Thousand Suns

If the radiance of a thousand suns were to burst into the sky,
that would be like the splendour of the Mighty One.

Bhagavad Gita

In late 1932 and early 1933 two events occurred only months apart,
which were to have devastating consequences. One was in politics;
the consequences of Hitler's rise to power were to dominate the
history of the remainder of the century. By 1939 most of Europe
was again at war. The second event was the discovery of the
neutron. This dominated the physics of the decade but attracted
little attention outside the, then small, international scientific
community. The cataclysmic implications of the discovery of the
neutron were not appreciated until the fateful year 1939. The rise
of extremism played a part in that delay. When it became clear
that the neutron was the key to unlock the massive amounts of
energy promised by Einstein's $E = mc^2$, physics and politics became
inextricably linked.

These two events dominated the development of the quantum
revolution in the 1930s and 1940s. Driven by the fear that scientists in Nazi Germany were working along similar lines the US
government funded the Manhattan Project to develop a nuclear
bomb in 1942. The most expensive scientific and technological
project hitherto undertaken was completed in almost complete
secrecy. Two very different approaches to the production of a
nuclear weapon were taken. Both succeeded in under three years.

A uranium weapon was used on Hiroshima and a plutonium weapon on Nagasaki; both were devastating.

A lesser-known consequence of these events was a delay in the start of the semiconductor revolution by 16 years. After the war ended, the relatively new industrial laboratories in the US could redirect their attention from the war effort. By looking back at the progress of the quantum revolution in 1931, they were able to kick-start the semiconductor revolution.

The entanglement of physics and politics that resulted from these events has had important ramifications for the development of the solar revolution. As we will see, President Eisenhower's description of civil nuclear power as 'Atoms for Peace' was something of a misnomer. Civil and military nuclear power were linked in the post-war years. This has had consequences, which have delayed the solar revolution in some countries.

The discovery of the neutron

The neutron, one of the villains of our story, was discovered in 1932. It was found as a result of over three decades of extensive study of *radioactivity*, the radiation emitted when unstable atomic nuclei decay. Nuclear radiation has re-emerged as a matter of public concern in 2011 following the disaster at Fukushima in Japan.

We saw in Chapter 1 that all the heavy nuclei in the universe were produced in the death-throes of first-generation stars. Most heavy nuclei are *stable* and have lasted for billions of years. They were swept up by our sun to form our solar system. Most of the *unstable* nuclei produced at that time decayed away in the first million years or so, including plutonium. Just a few very heavy and very long-lived unstable nuclei, left over from the death of the first-generation stars, lurked in minerals in the bowels of the earth. The heat energy generated in their decay helps to keep the earth's core hot enough to remain molten. As we will see, this geothermal energy can be exploited safely.

The radiations that are emitted when these unstable nuclei decay were identified through painstaking and dangerous research. This was led at the end of the nineteenth century by the French physicist Henri Becquerel and two French chemists, the husband and wife team of Marie and Pierre Curie. All three were awarded the 1903 Nobel Prize in Physics. Marie Curie was also awarded a Nobel Prize in Chemistry. She was the first person to receive two Nobel Prizes. Marie was born in Warsaw and one of her discoveries was a new radioactive element, which was named polonium after her native country.

The researchers discovered three main types of radiation. *Alpha* radiation consists of the very stable nuclei of helium atoms, *beta* radiation is formed of electrons and *gamma* radiation is due to electromagnetic waves like sunlight but with much higher energy and much shorter wavelength. The three types of radioactivity have very different properties. Each has their own particular health risks if not handled properly. Marie Curie died from leukaemia.

Alpha particles were used by Rutherford in his experiments to bombard gold and other heavy atoms. These were the experiments that established that the nucleus was a very small and very heavy component at the centre of all atoms. They inspired Bohr to come up with the first quantum picture of the atom. In 1931 German and French researchers were using alpha particles to bombard light nuclei when they observed a new radiation, which had different properties to alpha, beta and gamma radiation. The correct identity of this new radiation was established by the English physicist, James Chadwick. He had worked with Rutherford in the exciting Manchester days and was now working again with him in Cambridge.

Chadwick showed that this new radiation consisted of a particle with mass very close to that of the proton. Unlike the proton it had no electric charge, so it seemed appropriate to call it a *neutron.*

He showed the neutron had mass similar to the proton by placing a block of solid paraffin in the path of this new radiation. The neutron knocked protons out of the paraffin which were much easier to detect.

His experiment can be compared with a phenomenon well known to skilled snooker players. If you carefully propel the cue ball so that it hits a second, stationary ball dead centre, the first ball comes to a complete stop and the second ball takes on the speed and direction of the cue ball. This clever trick only works because both snooker balls have the same mass – a phenomenon well known to skilled physics undergraduates. Chadwick argued that the protons ejected from the paraffin had taken on the energy of motion of a new particle with very similar mass to the proton, just like the second snooker ball takes on the speed and direction of the first.

The discovery of the neutron, for which Chadwick received the 1935 Nobel Prize in Physics, solved a mystery which had perplexed physicists for two decades. Rutherford had shown there is an extremely small, heavy nucleus at the centre of all atoms. If the nucleus is made up of protons, they will repel each other. So how come most nuclei last for billions of years? I call the neutron a villain, but it does have some redeeming features. It is found in all nuclei apart from the nucleus of hydrogen, which is just a single proton. Also, there is an extremely strong attractive force between neutrons and protons, which is much stronger than the force of repulsion between protons. This force is so strong that most nuclei hold together for billions of years. Physicists have given this force an accurate, but somewhat unoriginal, name: the *strong nuclear force*.

As well as being enormously strong, the force only acts over the minute diameter of a nucleus. This helps to explain why it took so long to find direct evidence for this force. It soon became clear that this strong attractive force was a new form of quantum bonding. Remember from the previous chapter that the helium atom is very stable because it has two electrons in the first electron orbit. The nucleus of the helium atom, with two protons and two neutrons, is also a very stable nucleus. It is in fact the alpha particle, which Becquerel and the Curies had discovered is emitted when heavy nuclei decay.

In the early 1930s most people were nowhere near as excited by the discovery of the neutron as the physicists, nor were they aware of its possible dangers. The Great Depression and the rise of totalitarianism were quite enough to worry about. Also, what practical use could the neutron have? After all, in 1933, Rutherford, by now the grand old man of nuclear physics, famously dismissed the idea of using nuclear energy as 'talking moonshine'. It was not until the traumatic year 1939 that a small number of physicists realised that $E = mc^2$ energy could be unlocked by a neutron.

Physics and politics become entangled

Why did it take six years for physicists to appreciate the destructive potential of the new particle? In his fascinating and definitive account of the history of the nuclear bomb *Brighter than a Thousand Suns*, Robert Jungk describes how the rise of totalitarianism was incompatible with the objective thinking and free international collaboration on which science depends. A number of the leading figures of the quantum revolution in Europe were Jewish. The rising intolerance they experienced in German universities forced many to emigrate to the US, including Einstein himself.

Fascism was also on the rise in Italy where important experiments were underway in Rome led by Enrico Fermi. The Rome-born physicist had already made a number of theoretical discoveries in quantum mechanics. The electron and proton come from a family of particles called *fermions* in his honour. Now at the age of 33 his team in Rome was making important experimental discoveries. In 1934 they tried bombarding uranium, the heaviest stable nucleus, with the new neutron particle. The team became convinced that they had produced a new nucleus that was heavier than uranium. A new element heavier than any previously known element would be the ultimate alchemy.

Jungk relates how, during the period in which Italy was invading Abyssinia, the atmosphere in the institute was not conducive to scientific self-criticism. The preoccupation with the possibility that they had produced a new element had blinded them to the reality of the revolutionary nature of the nuclear debris left behind in their experiments.

It was not only the rise of fascism that delayed the correct interpretation of Fermi's experiments for four years. A second factor was the complexity of the chemical analysis necessary to identify the nuclei in the experimental debris. A third factor might well have been that the first person to suggest the correct answer and two of the leading experts in the analysis of the debris were all women. In an even more male-dominated scientific environment than today, this could have prejudiced researchers against the alternative explanation that two of the women were proposing.

A German chemist Ida Noddock, who had herself discovered the element rhenium at a young age, was one of the few experts who disputed Fermi's claim to have produced a new nucleus heavier than uranium. As early as 1934 she suggested that '... when heavy nuclei are bombarded with neutrons the nuclei in question might break into a number of larger pieces ...'

Meanwhile, the nuclear debris from Fermi's bombardment of uranium was being analysed in Berlin by other experts, the German Otto Hahn and an Austrian woman Lise Meitner. They both initially shared Fermi's view that a nucleus heavier than uranium had been formed. Jungk describes many disagreements between Lise Meitner and a French expert, Irene Joliot-Curie, daughter of Marie Curie, who favoured Ida Noddock's interpretation.

BRIGHTER THAN A THOUSAND SUNS

In late 1938, after Lise Meitner, who was a quarter Jewish, had been forced to find refuge in Stockholm, Fritz Strassmann, her successor as Hahn's assistant, at last persuaded his boss to change his mind about what was in the debris of Fermi's experiments. The account in Jungk's book of this pivotal moment in the history of science and civilisation is both fascinating and revealing.

Strassmann came rushing into Hahn's office with a copy of Irene Joliot-Curie's latest paper imploring him to read it. Hahn's initial response was 'I'm not interested in our lady friend's latest writings' while continuing to puff on his cigar. Strassmann persisted, summarising the crucial points in her analysis. Then, suddenly, Hahn himself got the message at last. How could they have overlooked what was so clear to Joliot-Curie? He rushed from the room with Strassmann to initiate a careful review of their data in the laboratory, leaving the cigar, unfinished, glowing on the desk.

Hahn and Strassman's experiments provided clear evidence that the nuclear debris from Fermi's bombardment of uranium contained the nucleus of the element barium, which has just over half the number of protons and half the mass of the original uranium nucleus. Fermi's neutrons had split the uranium nucleus in half as Ida Noddock had predicted.

The shutters were coming down all over Europe, but, as an act of defiance to the Nazi regime, Hahn sent a pre-publication copy of the paper containing these results to his old colleague Lise Meitner in Stockholm. She immediately realised that they had both been working along the wrong lines and that Joliot-Curie and Noddock were right. She wrote a paper with her nephew, Otto Frisch, in which they conjectured that on gaining a neutron the spherical uranium nucleus would form a waist, as if tightening its belt, before splitting into two nearly equal parts. This is rather like a biological cell dividing. Hence they called the new effect *nuclear fission*.

Hahn received the Nobel Prize for his work. The supreme scientific prize was not extended to Ida Noddock, Irene Joliot-Curie,

Lise Meitner or indeed Fritz Strassmann, who had all made important contributions.

Fermi had in fact 'split the atom' as early as 1934. It had taken nearly five years to confirm this partly because the analysis was difficult, partly because of the rise of totalitarianism but also because of scientific and cultural prejudices. Jungk has a most appropriate analogy for the mindset of the scientists who for many years thought that only bombardment at the highest energy could break the quantum bonds of the nucleus. It is like the mindset among many politicians and many of the military who believe that nuclear weapons are necessary for our protection. This preoccupation has blinded them to the uselessness of nuclear weapons in the face of a single suicide bomber who can penetrate our defences and cause devastation. *Just one single, low energy neutron can destroy a heavy uranium nucleus.* It is a tragically prophetic simile, which is even more appropriate today than when it appeared as a footnote to Jungk's book in 1958.

Jungk recounts the considerable difficulty a group of central European émigré physicists, led by the Hungarian Leo Szilard, found in warning President Roosevelt that a powerful nuclear bomb could be made and that Hitler might get there first, even though they had Einstein's support. The task of warning Churchill in the United Kingdom also fell to two émigrés: Otto Frisch, Lise Meitner's nephew, and Rudolph Peierls. After the war both became highly respected academics at Cambridge and Birmingham Universities respectively. As an undergraduate at Birmingham, I was extremely fortunate to receive excellent applied mathematics lectures from Professor Peierls.

It is ironic that Szilard had such trouble in getting his frightening message over to politicians. Since the Second World War, scientists have received a warm welcome from powerful politicians if their work has possible military applications. We will meet many examples in this book. It will be a recurring theme that one reason why solar technologies have yet to fulfil their potential is that the military have shown little interest in them. If the military are keen on a particular line of research, then, nowadays, so are the most powerful politicians.

The Manhattan Project

In just three years, from 1942 to 1945, massive scientific, technical and logistical problems were solved by the Manhattan Project supported by a huge financial commitment from the US government.

To appreciate the magnitude of the technical achievement we need to understand a little more about nuclear physics. Soon after Hahn's discovery that Fermi had split the uranium nucleus in half, Frederic Joliot-Curie, husband of Marie Curie's daughter Irene, came up with another important observation. Early in 1939 he showed that when a slow neutron shatters a uranium nucleus, the debris not only contains two nuclei of approximately half the size, but also *two or three more neutrons*. The clear possibility existed that, in an appropriately designed lump of uranium, these neutrons would split other uranium nuclei yielding even more neutrons. This would set up a *chain reaction*. Perhaps it could lead to a nuclear explosion?

Neutrons are present in all atomic nuclei, apart from the nuclei of hydrogen. They have been there providing the glue that holds the nuclei together ever since they were produced more than five billion years ago in the death of first-generation stars. It is not easy to work out how many neutrons there are in any one nucleus as they are neutral particles and have no effect on the negatively charged electrons flying around in their atomic orbits. It is the number of protons in the nucleus that fixes the number of electrons in orbit. All the neutrons do in a stable nucleus is hold the nucleus together and make it heavier.

Two nuclei with the same number of protons but a *different number of neutrons* are called two *isotopes* of the same nucleus.

THE HISTORY OF THE SEMICONDUCTOR REVOLUTION

As the two nuclei have the same number of protons, they both form atoms with the same number of electrons. Chapter 3 explained that the number of electrons determines how that atom bonds with other atoms. Therefore two isotopes of the same element form exactly the same molecules with other atoms. Hence the two isotopes cannot be separated by chemical reactions.

Here is one example, which will be important later in my story, of two atoms that differ only by the number of neutrons in their nuclei and so participate in the same chemical reactions. There is an isotope of hydrogen called tritium. Hydrogen, you may remember, has the simplest nucleus. All it consists of is a single proton. The tritium nucleus has one proton and two neutrons. One proton and one neutron would bind strongly together. In fact, they form the appropriately named, and very stable, deuterium nucleus. However, the extra neutron makes the tritium nucleus unstable. Over a number of years it decays, emitting beta radiation.

Tritium is a gas like hydrogen. It is emitted by nuclear reactors from time to time. Just like hydrogen it combines with oxygen to form molecules of water. Water containing tritium behaves just like ordinary water except that it is radioactive. Some radiation experts think tritium may be responsible for the increased risk of childhood leukaemia within 5 km of a nuclear reactor observed in a number of countries.

Returning to 1942, it turned out that only one isotope of the uranium nucleus – uranium-235 – can be readily split in half by neutrons. The number 235 refers to the total number of protons and neutrons. The uranium found naturally in nature mostly consists of the isotope uranium-238. One slow neutron that is captured by a nucleus of uranium-238 will not split the nucleus. The captured neutron will, however, turn the uranium-238 nucleus into an isotope of plutonium, namely plutonium-239. This is what Fermi and his team thought, erroneously, had occurred in their early experiments in Rome.

A massive factory was quickly built in Oak Ridge, Tennessee to separate out the small amount of uranium-235 isotope from the much more plentiful uranium-238 in natural uranium. The two isotopes cannot be separated by chemical reactions as they have the same number of electrons, but they can be separated by physical means on the basis of the very small (1 per cent) difference in their nuclear masses. A gas made from the isotope uranium-235 will pass through a membrane slightly faster than the same gas made from the heavier isotope uranium-238. By repeatedly passing the gas through membranes, the gas eventually contained more of the lighter isotope; this process is known as the *enrichment* of uranium-235.

As an insurance policy, should this large factory involving many complicated stages not work, a smaller, but even more problematic system was devised to produce significant amounts of plutonium-239 isotopes. A team led by Enrico Fermi, by then at the University of Chicago, constructed the world's first nuclear reactor beneath an old squash court on campus. Neutrons produced by the splitting of uranium-235 were slowed by blocks of graphite and captured by the uranium-238 nuclei to form plutonium-239. Fermi had produced the first controlled nuclear fission reaction.

In one of the great ironies of science, Fermi achieved in Chicago in 1942 what had erroneously preoccupied his team in Rome in 1934. It is a sobering thought that had they guessed correctly eight years earlier, fascism might have triumphed in the Second World War.

Both the uranium enrichment and the plutonium production approaches worked, but too late for the bomb to be used in the war in Europe. Uranium was used in the bomb that was dropped on Hiroshima. Plutonium was used in the bomb dropped on Nagasaki.

This practical demonstration of the quantum revolution was infinitely more problematic, complex and more expensive than the discovery of semiconductors, which followed a few years later. It was driven by the very real fear that Germany was developing a similar weapon. There were still talented and knowledgeable scientists such as Heisenberg in Germany. Many feared that Hitler's decisions to take on Western Europe, the USSR and the US were so irrational he must have a super-weapon in reserve.

Only a small fraction of the 150,000 people on the Manhattan Project knew the objective of their work. For those that were aware, the moral dilemma was immense. Should the bomb be used? If so, on what target and with what warning? The dilemma became more intense in 1945 as it became clearer that war with Germany would end before the bomb was tested. Ironically, Leo Szilard, who had successfully alerted Roosevelt to the possibility that Germany was developing a bomb, now found himself leading a group of scientists lobbying Roosevelt not to use the bomb on Japan. They correctly feared it would make it more difficult to obtain an international agreement to prevent a nuclear arms race after the war. Jungk describes how a letter from Szilard, and a supporting letter from Einstein, were lying on the presidential desk, untouched, when, on 12 April 1945, Roosevelt died unexpectedly.

Joseph Rotblat, a Polish émigré physicist working on the Manhattan Project, took a courageous independent stand. As soon as it became clear that Hitler was not working on the bomb, he left the project to work on medical applications of the new physics. After the war he founded Pugwash, a movement of scientists dedicated to halting the nuclear arms race. Einstein was a co-founder of Pugwash. He regretted his support for Szilard's attempts to alert Roosevelt in 1939. Jungk quotes Einstein as saying 'If I had known that the Germans would not succeed in constructing the atom bomb I would never have lifted a finger.'

Jungk describes the moral dilemmas felt by many of the scientists involved. Most memorably he recounts the thoughts of the scientific leader of the project, Robert Oppenheimer, at 5.30 a.m. on 16 July 1945. This was the time of the first, and only, test before either type of bomb was used in anger. The test, code named Trinity, took place in Alamogordo, New Mexico. Inside the control room, over five miles from the detonation point, Oppenheimer was clinging to a pillar reciting the words from the Bhagavad Gita:

> If the radiance of a thousand suns were to burst into the sky,
> that would be like the splendour of the Mighty One.

A dazzling white flash was followed by a bright, growing ball of flame. This in turn was followed by the sinister mushroom cloud which epitomised the new technology. Shocked, Oppenheimer was reminded of another line from the same source:

> I am become Death, the shatterer of worlds.

The first nuclear weapon, the uranium one, was exploded without warning over the city of Hiroshima on 4 August 1945. The second one, a plutonium weapon, was exploded over the city of Nagasaki five days later. Japan surrendered unconditionally on 15 August. Oppenheimer was one of four scientists on a panel that advised President Truman on the use of the bomb. They could have advised a demonstration on a remote site. They might have done so, according to Jungk, if they had been aware that Japan was putting out peace feelers.

What became of the Manhattan scientists?

I have dwelt on the story of the Manhattan Project not just because it was the first large-scale application of quantum ideas. It was also the first example of a massive technological project that achieved its aims in a relatively short time due to unlimited government backing because of a real or perceived threat. Both routes to a nuclear bomb were greater scientific and technological achievements than harnessing solar power. Despite the existential threats our civilisation now faces, many governments appear unwilling to act, for reasons I will explore in this and later chapters.

After the war ended many Manhattan scientists and technologists moved in one of three directions. Some moved on to the fusion bomb project, which produced the even more powerful hydrogen bomb under the leadership of the Hungarian émigré physicist Edward Teller. Again the technological challenge was immense: they harnessed the explosive energy of a primary fission bomb to compress a secondary target so fiercely it reproduced the conditions at the centre of the sun. Again it was achieved relatively quickly with unlimited government backing. In this case, the motivation was the fear that their former allies in the USSR might get there first.

Others moved back to academia. The most notable was Oppenheimer himself who in October 1945 unexpectedly resigned as director of the Los Alamos laboratory. He retained influence with the US government as a consultant until 1954. In April of that year, at the height of the cold war, he was stripped of his security clearance because of his early association with communists and alleged opposition to the fusion bomb programme.

A number of Manhattan scientists followed Oppenheimer back into academic research. Many of them used their knowledge of quantum theory in new areas of particle physics and astrophysics research. Some, including Richard Feynman whom we will meet again later, found new ways to link quantum ideas to Maxwell's theory of electromagnetism.

Other Manhattan scientists moved in a third direction, which became known as 'Atoms for Peace'. They were motivated by the wish to see the plutonium-producing reactors of the Manhattan Project put to more peaceful ends such as generating electricity.

The origins of civil nuclear power in the UK

Who could possibly object to an ideal like 'Atoms for Peace'? I believe that most of the scientists and technologists who decided to work on the peaceful applications of the new energy resource were motivated by sentiments similar to those in the Book of Isaiah,

'They shall beat their swords into ploughshares, and their spears into pruning hooks.' In debates I have been asked, in somewhat different words, 'surely, a muesli-eating, sandal-wearing, aging hippie like you is in favour of beating the nuclear sword into an electric ploughshare?'

I am not. Sadly, any blacksmith skilled in the art can beat a ploughshare back into a sword. In the case of the UK, unscrupulous members of the political and military establishment got skilled nuclear engineers to do just that. How they did so is not simply a matter of historical curiosity. Military requirements dictated the design of the first generation of UK civil nuclear reactors, influenced the design of the second generation and locked the UK into a fixation with reprocessing nuclear waste. The sorry tale also explains why the UK is now left with the dangerous embarrassment of the world's largest stockpile of separated civil plutonium. This, as we will see, may be dictating the need for a whole new generation of nuclear reactors.

In 1956 the nation saw images of a young Queen Elizabeth connecting the world's first electricity-generating nuclear power station at Calder Hall to the national grid. This electricity was in fact a by-product of the main function of those reactors, which was to produce plutonium for the nuclear weapons programme.

Weapons designers prefer the isotope plutonium-239 for effective bombs. Plutonium that is mostly plutonium-239 is known as *weapons-grade* plutonium. Plutonium-239 is the first isotope of plutonium produced when neutrons bombard uranium nuclei in a reactor. The military reactors at Calder Hall could only be run a few months at a time. Then the low-irradiated fuel had to be removed before other isotopes of plutonium formed. Only a little plutonium was produced in that time, but it was mainly plutonium-239.

Frequent shutdowns for refuelling are not ideal when supplying electricity to the grid. The UK's first civil nuclear stations were larger versions of the Calder Hall reactors. To avoid these shutdowns, they were designed with an important new feature: the facility to remove and replace some of the irradiated fuel while the reactor was running. Hence electricity could be continuously generated. Replacement of some of the fuel would start early in a reactor's life. This ensured the reactor did not reach a state where all the fuel needed replacing at the same time.

In the early operation of the first British civil reactors, significant amounts of plutonium-239 were produced and extracted. Just as electricity was a by-product of the military Calder Hall reactors, plutonium-239 was a by-product of electricity generation in the first civil stations.

In the late 1950s, beating ploughshares back into swords in this way seemed a convenient arrangement. However, the scheme had profound political and technological implications for civil nuclear power in the UK.

Plutonium in low-irradiated fuel is of no use to the military until the fuel has been *reprocessed*. This is a complex, costly and potentially dangerous chemical procedure in which plutonium and uranium are separated from the irradiated fuel. A reprocessing line was constructed at the plant now called Sellafield to take all the irradiated fuel from the first generation of civil reactors and the military plutonium reactors. Low-irradiated fuel from the civil reactors was reprocessed in this line at the same time as low-irradiated fuel from the Calder Hall military reactors. This is how the weapons grade plutonium was separated.

The UK civil nuclear industry became locked into reprocessing spent fuel, unlike many other countries, because of the demand for weapons grade plutonium. When the second generation of UK civil reactors came into operation a new reprocessing line was built though the plutonium in the new reactors was no longer useful for weapons. The rational for this second reprocessing line was that it was environmentally advantageous, a foreign exchange money-spinner and a fast breeder reactor would be built to burn the plutonium. None of these hopes were fulfilled.

Britain has ended up with the world's largest stockpile of separated plutonium, over 100 tonnes. There are also large amounts of other dangerous waste products. The temptation this plutonium provides for terrorists is now used as an argument for building a third generation of nuclear reactors in which to burn the plutonium. However, shipping the plutonium around the country to reactors will increase the opportunities for terrorists. Furthermore, it only postpones the problem of disposal. New reactors will produce new plutonium.

What happened to the plutonium produced by the first civil reactors when they stated up in the 1960s? Early in that decade, the British ballistic missile programme ran into major difficulties and had to be scrapped. The UK government therefore came to an agreement with the US for the supply of Polaris nuclear missiles. As part of the deal Britain agreed to provide America with plutonium. They expected to have a copious supply when the civil reactors started up in the mid-1960s.

As the decade passed, the British side of the bargain unravelled. First, the mechanism for refuelling the civil reactors while they were working proved problematic. This resulted in delays in supplying the plutonium to the US. Second, another branch of the British government, the Foreign Office, possibly unaware of what their colleagues in the Ministry of Defence and Ministry of Power were up to, started promoting the Non-Proliferation Treaty (NPT). Tony Benn, who was the minister responsible for the UK civil reactors in the late 1960s, was not made aware that plutonium from the civil programme was being sent to the USA while he was in charge.

The UK is one of the original three 'depositary states' for the articles of the NPT. The basis of the treaty is the separation of civil and military nuclear activities. Hence the British government's promotion of the NPT was a classic case of 'do what we say, not what we do'. The treaty came into force on 5 March 1970. Records show that for a few years prior to this, two of the civil reactors, at Hinkley Point in Somerset, were refuelled at great speed; far faster than was optimum if civil electricity generation was the objective.

In the end the plutonium consignments to the US mainly contained high-irradiation civil plutonium that was used in research reactors. But in return the British government has received nuclear material from the US which has clearly benefitted the UK military nuclear programme.

Unravelling the history of nuclear ploughshares

Some details of the nuclear exchanges came to light in the early 1980s. This was mainly due to the efforts of a whistle-blower, the late Ross Hesketh, who worked at a civil nuclear laboratory run by the electricity generating board. In response, in 1983, the government provided some limited information about the exchange with the US. However, they also made categorical assurances that none of the civil plutonium had ever gone into weapons.

Some colleagues and I decided to test these assurances. We calculated from first principles the amount of plutonium produced in the UK civil reactors. The data required for this analysis was difficult to obtain. We were helped by the CND Sizewell Working Group. CND was an objector to the building of the Sizewell B nuclear reactor during the public inquiry, which ran from 1983 to 1985.

Cross-examination by CND of nuclear industry spokespersons at the Sizewell Inquiry uncovered a pathway for civil plutonium to leak into the military stockpile. British Nuclear Fuels Limited (BNFL), the government-owned company that operated Sellafield, admitted that low-irradiated fuel from both military and civilian reactors was reprocessed at the same time in the same reprocessing line. Indeed, the practice at Sellafield was to refer to low-irradiated plutonium from both civil and military reactors as 'military' plutonium. Under further cross-examination, BNFL admitted that the relevant safeguard authorities were concerned about this co-processing of civil and military low-irradiated fuel. It meant that they were not able to verify the amounts of civil plutonium being reprocessed.

In 1985 we published our calculations in the journal *Nature*. These showed that more plutonium had been produced in the civil reactors than was in the locations admitted by the government. In April 1986, in response to the revelations CND extracted at the Sizewell Public Inquiry and, I like to think, to our paper, Margaret Thatcher's government made one of its rare U-turns. They added a significant caveat to their previous assurances, namely that civil plutonium had not been put to military use 'during the period of this administration'. The earlier government assurances that civil plutonium had not gone into weapons no longer applied to the years in which the civil reactors were producing weapons-grade plutonium.

The confirmation that the nuclear ploughshares had been beaten back into swords came in 2000. As part of its Strategic Defence Review, the new Labour government decided to publish an inventory of British military plutonium. The Clinton administration had published its own inventory of American military plutonium a few years earlier. As it happened the US data agreed with our 1985 figure for the amount of UK civil plutonium sent to America, when adjusted for new information on plutonium in waste.

The UK review came up with a remarkable admission in 2000. The UK had more weapons-grade plutonium than it thought it had. In the words of the Ministry of Defence report, 'the weapon cycle stockpile is in fact some 0.3 tonnes larger than the amount of plutonium the records indicate as available.'

The entry in the data tables admits that 0.37 tonnes of weapons-grade plutonium came from 'unidentified sites'. The report does not attempt to identify the origins of this 70 bombs' worth of plutonium despite there being only one possible source of that much weapons-grade plutonium. The origin must have been the low-irradiated fuel from the civil programme that was processed together with the military fuel at Sellafield.

This is confirmed by our 1985 *Nature* publication. Our calculations showed that 0.36 tonnes of weapons-grade plutonium were produced in the early years of operation of the civil reactors. This is remarkably close to the 0.37 tonnes the Ministry of Defence identified as coming from 'unidentified sites'.

Is history repeating itself?

I have gone into detail on this history because it goes some way to explaining the preoccupation with nuclear power in the political, military and scientific establishments in the UK. This, in turn, has meant that Britain has fallen behind counties like Germany and Italy in taking up solar power as an alternative to nuclear. It also explains why Britain finds itself with a dangerous stockpile of over 100 tonnes of separated plutonium. This stockpile is now being used as a justification for building a whole new generation of nuclear reactors.

Could history be repeating itself? Could there again be a military imperative behind the next UK civil nuclear programme? I doubt that the supply of weapons-grade plutonium is a factor this time round. Plutonium-239 can be recycled from old warheads to new ones.

Two other nuclear materials important to the weapons programme do need constant topping-up. Radioactive tritium, which we encountered earlier in this chapter, is a vital component of nuclear warheads. Second, highly enriched uranium is the fuel that powers the Trident submarines and the hunter-killer submarines. The main source of supply for these materials is probably the US under the defence agreement referred to earlier. Tritium is produced in reactors and low-enriched uranium is an important part of the civil nuclear fuel cycle. Possibly the military and political establishment see a new programme of nuclear reactors as a fall-back for supplying these important items should relations with the US cool.

Surely international safeguards would prevent Britain from obtaining tritium or further enriching uranium from the civil nuclear programme? Sadly, the UK has set an unhappy precedent by removing safeguarded nuclear material from the civil programme for military activities. This is allowed if due notice is given to the safeguards authorities. In the 1980s vast amounts of depleted uranium (the waste left behind when uranium is enriched) belonging to the UK civil programme were removed from safeguards. Some was used to produce nuclear material, probably tritium for warheads. The depleted uranium was also used for hardening shells and strengthening tank armour. The investigative journalist John Pilger has highlighted the human cost of the radioactive debris from the use of UK depleted uranium in Iraq. Though depleted uranium is not as radio-active as enriched uranium, a suggestion has been made, which has yet to be experimentally verified, as to why it might be genetically damaging.

Another factor in the enthusiasm of successive British govern-ments for new nuclear reactors could be that a new programme will require an enhanced supply of nuclear engineers from the universities. Their expertise would also be useful for engineering warheads and the nuclear reactors that propel submarines.

I suspect, however, that one of the UK government's main moti-vations is a belief that only with a new nuclear programme will the UK be able to take its place, alongside the US and France, in exerting control over a future 'plutonium economy'. They quite rightly want to ensure it is run in a way that avoids nuclear weapon proliferation and terrorist outrages. My personal view is that, given Britain's poor record on plutonium accountancy and the military use of depleted uranium from the civil programme, we are hardly the best country to set standards for safeguarding international trade in plutonium.

We will find in Part II that all our electricity demand can be safely and cheaply derived from the renewables, so a 'plutonium economy' is completely unnecessary. Rather than using our dangerously large plutonium stockpile as a reason for a new generation of nuclear reactors, the UK nuclear industry should concentrate on spending the £3 billion a year to which the government is committed, to develop a safe way to dispose of the plutonium stockpile. Remember, it must be kept out of the environment and out of the hands of terrorists for more than 300,000 years.

I suspect that all these factors play some part in the inordinate determination of successive British governments to ensure new nuclear reactors will be built. As costs have soared they have moved from a position of 'allowing the market' to invest in new-build to the current government's position of committing vast amounts of taxpayers' money to rig the electricity market in favour of nuclear with 'contracts for difference' to guarantee a high price for nuclear electricity. It is still not clear that this policy will convince 'the market' to invest the massive sums required for a new generation of nuclear reactors. As we will see, the crunch will come when one or other of the prototypes eventually work. There is always a lot to learn from a prototype, particularly one as big as a nuclear reactor. Also, the Fukushima disaster has led to major design changes.

Are 'Atoms for Peace' working in other countries?

It was hard work studying the extent of the overlap between civil and military nuclear activities in the UK as much information was covered by 'national security'. I am certainly not in a position to comment on how important the link is in other countries with nuclear weapons. What is clear is the extent to which all the countries with nuclear weapons have lagged behind in the solar revolution.

Let's look at data on PV power installations from the International Energy Agency, Photovoltaic Power Systems (PVPS) programme.

By the end of 2011 the three leading countries in terms of PV power installed were: Germany, Italy and Japan. Can you think of any historical similarities between these three countries? They were all defeated in the Second World War. There was no way the victorious allies would allow any of these three countries to develop nuclear weapons. Clearly Japan had two very good reasons for not wanting to have anything more to do with such weapons.

I am very encouraged that these three countries, which were once profoundly undemocratic, now lead the way in implementing PV. Relevant decisions were taken democratically in Germany and Italy. In all three countries, individual members of the public have invested large amounts of their own money to help implement PV. The Italians have had two referenda, one that closed their civil nuclear programme and a more recent one that decided against any new nuclear projects. The German PV programme was introduced by the Green–Socialist coalition government along with a decision to not to replace aging nuclear reactors, as we will see in Part II.

The Japanese situation is more complicated. In the mid-1990s, the Japanese government introduced a 70,000-roof PV programme. This started well before the successful German 100,000-roof programme. Japan led Germany in total PV installations until 2005, when Germany's steady exponential rise and faltering Japanese government support for PV produced a cross-over point. I well recall my Japanese colleague's dismay at their government decision to prioritise nuclear over PV around that time. It is a decision Japan must have regretted when the Fukushima disaster occurred in 2011. Also there must have been regrets when, in 2012, their old rivals China caught up with them in PV installations.

Can it be a coincidence that Germany, Italy and Japan, three countries with many reasons for not having nuclear weapons, were leading the solar revolution in 2011? Let's look in more detail at the PVPS data. Five countries out of the 23 are nuclear weapon states. What are the positions of these five countries in the PV league table? By the end of 2011, Germany had installed six times as much PV as the US and eight times as much as China, despite being a much smaller country than either. Furthermore, Germany had installed 9 times as much PV as France and 25 times as much as the UK. At least the UK has shown some improvement. In 2010, Germany had 250 times more PV than Britain. Finally, Germany had 131 times as much PV as Israel in 2011, though the latter is a much sunnier country. In the words of courtroom drama: I rest my case.

Winding back the clock

We have followed the 'Atoms for Peace' trail from the Second World War through to the present day to try to understand why the solar revolution is taking off at such different rates in different countries. In doing so, I have got ahead of myself in explaining how the quantum revolution turned into the semiconductor revolution. As early as 1931, a quantum theory of semiconductors had been suggested, but little practical work was done. In the 1930s and throughout the Second World War, quantum physics research was dominated by the aftermath of the discovery of the neutron and the exploitation of $E = mc^2$.

Now we must wind back the clock to the immediate post-war period. We will follow the semiconductor revolution, which developed very differently from 'Atoms for Peace'. We will find, though, that the new semiconductor technology developed most quickly when governments and the military were most interested in the application.

FIVE

The Mystery of the Quantum Conductor

We have now explained a remarkable mystery – how an electron
in a crystal can ride right through the crystal and flow perfectly
freely even though it has hit all the atoms.

Richard Feynman

The Second World War ended with much of Europe and Asia
devastated. Any surviving industrial and academic research was
geared towards the war effort. By contrast, much of US industrial activity had been stimulated by the demands of the war
and a number of strong industrial laboratories had emerged.
Their major interest was electronics for communications. It was
hardly surprising that, with the arrival of peace, the next major
step in the quantum revolution took place at the Bell Telephone
Laboratories in New Jersey.

This important development was the practical demonstration of electronic devices made from a novel form of electrical
material: the *semiconductor*. Quantum mechanics is crucial to
understanding how semiconductor crystals work. To understand
the revolutionary nature of semiconductors, just think about the
technology they replaced.

Electronics had made a major impact in the Second World
War. It facilitated wireless communication with aircraft and
between mobile armoured divisions. It helped radar waves detect

enemy bombers at night. Wartime electronics was based on more advanced versions of the diode and triode valves, which we met in Chapter 2. Fleming and de Forest's inventions, which had made Marconi's wireless a more practical proposition, had become more sophisticated during the Second World War.

After the war, these valves were used in the early computers which developed from the mechanical calculators used by wartime code-breakers. When peacetime manufacturing of consumer goods returned, valves made possible a post-war boom in domestic radio and television. Bulky wooden boxes of electronics spread through suburban living rooms in the 1950s.

Semiconductors swept all this technology away. The quantum revolution and, in particular, the wave and particle natures of the electron, were crucial to this transformation. Without quantum theory, the development of personal computers and modern communications technology would not have been possible. Quantum ideas helped unravel the solution to a mystery about electron behaviour, which was crucial for the development of the silicon chips that make all modern electronic gadgets work.

Like all the best mysteries the solution emerged where least expected; the clue was in the way electrons moved in the wires connecting the valves rather than the valves themselves.

Conductors and insulators

The electronics that helped win the war was based on a simple picture of electricity. It did not depend on quantum ideas. Valves were connected by metal wires, called *conductors*, through which electric currents flow. Materials in which electric currents *cannot* flow are called *insulators*. The flex of an electric kettle contains two or three copper wires, which are good conductors. Each wire is surrounded by a coloured plastic insulator, which stops current flowing from wire to wire. These coloured insulators are then covered in a white or black plastic insulator, which protects us from electric shocks.

Before quantum ideas changed everything, electricity was pictured as a flow of electrons in a wire like water in a pipe. An electric field in the wire forces the electrons to go one way. This is like gravity forcing a stream to flow downhill or a pump pushing hot water round our central heating systems.

I explained earlier how the electric field is generated a long way away by rotating wires in a magnetic field. The rotation is driven by hydropower, wind power or the burning of fossil or nuclear fuels. The electric field is transmitted in wires strung between tall pylons through many transformers to the high voltage contact in your electric socket. It then passes along one wire in the flex to the kettle where it pulls some electrons through the element to heat the water. Then the field returns through a second wire to the low voltage contact in the socket, back through transformers and over pylons to the generator. There are simpler ways to boil a kettle. Semiconductors are the key.

Sophisticated electronics based on valves linked by conducting wires helped win the war and made radio, television and the early computers work. Electronic engineers designed clever circuits without needing to know that electrons are quantum particles. It was physicists like Feynman who worried about why the current in the wire is the same as the current inside the valve. How is this possible, they asked? The valve contains a vacuum, but the wire contains atoms so tightly packed their electron orbits nearly touch. As a boy, growing up in New York City, Feynman liked fixing broken radio sets. This may be when he first decided the way electrons flow in wires was a 'remarkable mystery'.

The valves themselves were both cumbersome and delicate. Inside a fragile glass vacuum tube, an electric field between two metal contacts made electrons move one way. In a triode valve, a metal grid inserted between the contacts controlled the electron flow and so amplified radio signals. A valve needed a bulky power unit to provide the electric field and to supply the heater, which increased the electron flow. I remember how long it took for a radio to 'warm up'. You could tell which valve had 'blown' if its heater didn't glow.

The semiconductor revolution started in 1947 when scientists at Bell Laboratories in New Jersey demonstrated a device called a *transistor*, which acted like a valve but was made from a single crystal smaller than a postage stamp. Quantum ideas were crucial in explaining the mystery of how the crystal worked. The clues came from understanding how electrons move in conductors and why they don't move in insulators. Let's first look at the quantum picture of an insulator.

The quantum picture of an insulator

In Chapter 3 we saw how electron sharing leads to quantum bonding. These same ideas explain why insulators such as plastics are so different from metals such as copper. Beautiful and expensive diamond crystals make the best insulators, though clearly not the cheapest. I also find it amazing that diamond crystals, chimney soot and the graphite in pencils are, all three, made of carbon atoms. Diamond, a sparkling crystal, hard enough to cut glass, could hardly be more different physically to soot and graphite. Their differences can be explained by the way the quantum bonds hold the carbon atoms in different shapes. Quantum bonding also explains why diamond is a very good insulator.

Remember that a carbon atom has four electrons in its outer electron orbit and would like to have eight. If one atom of carbon is surrounded by four other carbon atoms it can share one of its four electrons with each of its neighbours. In return, each of the four neighbouring carbon atoms can offer up one of its four electrons to the original carbon atom. The original carbon atom now thinks it has a stable, full outer orbit of eight electrons.

Now comes the important part. Each of the four carbon neighbours, which share one of their electrons with our original carbon atom, can themselves, in turn, share their remaining three electrons with another three carbon atoms. Each carbon atom wants to share electrons equally with its four nearest neighbours and so needs to be equally distant from each of them. Nature achieves this by having the four neighbours at the four points of a tetrahedron, a shape that has four points and four triangular faces.

There is an educational toy that consists of magnetic balls and equal length magnetic rods that can be held together in different shapes. If you have played with this toy, you may have found by trial and error that the tetrahedron (six rods, four balls and four triangular faces) is the strongest shape. In diamonds, each carbon atom is at the centre of a tetrahedron, equally distant from its four neighbours. This pattern repeats throughout the diamond forming a strong and beautiful crystal.

This electron sharing between carbon atoms is very much like the bonding between pairs of hydrogen atoms to form a hydrogen molecule, which we saw in Chapter 3. Remember the problem of separating the two children when there were two seats left in musical chairs? Quantum bonding forms the very regular and very strong, three-dimensional arrangement of carbon atoms known as diamond. Physicists call such regular patterns of atoms *crystals*. Nature has many examples but few as simple, strong and as beautiful as diamond.

The sharing of the electrons in the outer orbits, as electron waves resonate from one atom to a neighbour and back again, holds the whole crystal together strongly. An electron can also resonate to other neighbouring atoms as long as it changes places with an electron on the neighbouring atom. When every atom has a full electron orbit, Pauli's exclusion principle is not violated if two electrons exchange places. In this way, electron waves spread between neighbouring atoms throughout the crystal. These spreading electrons form what is called a *band* of electron waves stretching across the crystal. Such a band, made up of all the electrons in full, outer orbits, which hold the crystal together, is called a *valence band*.

Why is diamond such a good insulator? An electron always remembers it is not only a wave but also a particle. Thanks to electron sharing, each carbon atom has a full complement of eight electrons in this outer orbit, which has formed a band throughout the crystal. If an electric field is applied an electron cannot jump to the next atom in the direction dictated by the field because the neighbour also has a full complement of eight electrons. Pauli's exclusion principle will not allow it to jump. There are no empty spaces in any of these outer atomic orbits for the electrons to jump into, so no current flows when an electric field is applied. Hence pure diamonds are excellent, but expensive, insulators.

The quantum picture of a conductor

Now let's see how an electron in a conductor behaves. A metal like copper is a good conductor. The outer electron orbit of a metal atom usually has one electron, though there is room in the orbit for eight. A single electron in an outer orbit is hyperactive, like a child seeking an empty seat in musical chairs.

The atoms in a metal are packed close together, with outer electron orbits nearly touching. As all the atoms in the metal are identical, an electron will have the same wavelength and frequency in the outer orbit of every atom. Atoms that are close together and on which electrons have the same frequency are ideal for an electron to resonate from one atom to the neighbour and back again. It is free to do so because there is room for eight electrons on both atoms. The neighbouring atom also has one electron in the same orbit, which can play the same game with the first atom. Hence pairs of metal atoms can share two electrons and bond rather like hydrogen molecules. The atoms in a metal are therefore held together. A crystal is formed, though it is not as strong as diamond where eight electrons are shared. A diamond will scratch most metals.

There is one crucial difference between the way electron waves behave in a metal and in a molecule like hydrogen or an insulator like diamond. Once an electron in a metal has jumped to a neighbouring atom, it need not jump straight back as it does when holding a hydrogen molecule together. There are many other identical atoms close by with nearly empty orbits on which the electron will have the same frequency. There is nothing to stop our original electron, as a wave, resonating with a nearby third atom and, as a particle, jumping to the nearly empty orbit on this third atom. The electron can then move on to a fourth one and so on throughout the crystal.

As a result of this jumping and resonating between atoms, there is an electron band in a metal crystal, like the valence band in diamond, but in this case formed by electrons in nearly empty orbits. Electrons can move freely in this band. If there is an electric field around, they will all move in the same direction and produce an electric current. Hence the band is called a *conduction band*. This is very different behaviour to electrons in an insulator. Remember electrons cannot move in response to an electric field if they are in the full valence band of an insulator.

Electrons move in the conduction band of a metal by jumping and resonating from one nearly empty electron orbit to another identical one in a nearby atom. This explains Feynman's 'remarkable mystery'. An electron can move in the conduction band of a metal crystal as freely as in a vacuum. Nearly empty, outer orbits of identical atoms packed close together are ideal for an electron particle to jump between, and for an electron wave to resonate between. The electrons ping their way through the crystal from one nearly empty outer orbit to another identical one. In Feynman's words, an electron *can ride right through the crystal and flow perfectly freely even though it has hit all the atoms*.

You may wonder why physicists describe metals as crystals

when a copper wire looks nothing like a diamond. Quantum ideas also explain this paradox. The electrons in the conduction band are free to move. They all have negative electric charge so they all repel each other. They try to get as far apart as they can, so they all end up on the surface of the metal. When sunlight falls on the surface, these electrons reflect the electromagnetic waves so that copper makes a good mirror.

On the other hand, sunlight can enter a diamond crystal as none of the electrons are free to move in this way. In diamond, the sunlight is refracted (not reflected) by the electrons bound to the atoms in the crystal to give the beautiful glinting effects on which people are tempted to spend outrageous sums.

What is a semiconductor?

In 1947, three Bell Laboratory physicists used these still relatively new quantum ideas to demonstrate a novel semiconductor device; the *transistor*. In terms of the worldwide impact of a scientific discovery, it must be up there with the major medical break-throughs like penicillin and antibiotics. Most people have at least a hazy idea how these medicines work. After all, we all have the personal experience of being cured by them. But explaining how a semiconductor works requires difficult quantum ideas.

Even the engineers who design the most complex of integrated circuits probably treat their semiconductor building blocks as tiny diodes or triodes without thought about the basic physics. Braun received the Nobel Prize for his contribution to the development of wireless without understanding that his rectifiers were semi-conductors. I doubt if many of the residents of Silicon Valley south of San Francisco understand the physics behind that name. But with some knowledge of quantum conductors and quantum insulators it is not so difficult to join the elite ranks of those who understand semiconductors, and solar cells for that matter.

THE MYSTERY OF THE QUANTUM CONDUCTOR

A semiconductor is a novel type of electric material – both insulator and conductor. It is a quantum insulator at everyday temperatures and a quantum conductor at very high temperatures.

Silicon is the most common semiconductor and the first hero of the semiconductor and solar revolutions. At normal temperatures, which we often refer to as *ambient* temperatures, a pure silicon crystal is an insulator like diamond. Like carbon, a silicon atom has four electrons in its outer electron orbit. By electron sharing with four neighbouring atoms, silicon forms a beautiful, strong, tetrahedral crystal – just like diamond. It has a full valence band holding the crystal together – just like diamond.

So why does silicon change from an insulator to a conductor at high temperatures? Every silicon atom has a number of electron orbits. The inner orbits are all full of electrons. The next orbit is the one with four electrons, which become eight in a silicon crystal thanks to electron sharing. There is an electron orbit at larger radius and higher energy but it is empty. There are no electrons in this more distant orbit at normal, everyday temperatures.

A silicon atom is like a house with a room on the ground floor suitable for eight children to play musical chairs and a playroom on the first floor where children can go when eliminated from the game. At everyday temperatures, all the children are engrossed in musical chairs on the ground floor; there are none in the first-floor playroom.

A silicon crystal, built up from many identical silicon atoms, is like a Silicon Street party organised at the local community centre. All the children are playing Silicon Street's favourite game, musical chairs, in the ground-floor playroom. They are contentedly seated in a large circle of chairs, one child per chair. This represents the valence band; the street party is a bonding experience for the Silicon Street community. A room on the first floor has been hired so children can let off steam when eliminated from the game. This first-floor playroom represents the electron band that would be formed from the next, empty electron orbit. If there were an electron in this band it would be free to move because the band would be nearly empty. As in a metal, this nearly empty electron band in a semiconductor is called a *conduction band*.

When all the children start playing musical chairs on the ground floor, the first-floor playroom is empty. The community centre represents the crystal of a perfect insulator. The two party rooms represent the electron bands. A pure silicon crystal is a perfect insulator at everyday temperatures; the valence band is full and the conduction band is empty.

A crucial feature of the community centre is the height of the first-floor playroom above the ground floor, or the number of steps on the stairway up from the ground to first floor. In a semiconductor, this important quantity is called the *band-gap energy*. It is the energy an electron needs to jump from the valence band to the conduction band and hence turn a silicon crystal from an insulator into a conductor.

In the next chapter we will discover that there is an amazing coincidence about the band-gap energy of silicon, which is of supreme importance for the solar revolution. It is also just one of a number of coincidences without which the semiconductor revolution itself would not have happened.

How large is the band-gap energy of silicon? It is very much larger than the typical heat energy that atoms and their electrons have at ambient temperature when they jiggle around their fixed positions randomly. There are many more steps from ground floor to playroom at the community centre than the energy of motion of the children jiggling impatiently in their seats downstairs.

The higher the temperature, the faster atoms jiggle and the hotter a crystal feels. At a very high temperature, the jiggling increases so much that the electron energy of motion is enough for some electrons to reach the conduction band just by jiggling. In the conduction band, they are free to move. This is why a semiconductor like silicon becomes a conductor at high temperature.

The temperature where this happens is too high to be useful for electronic devices. This means that silicon is an ideal insulator at room temperature. So what is revolutionary about semiconductors? In 1947, the pioneers discovered three new and revolutionary ways in which a crystal insulator like silicon could be turned into a conductor *at normal, everyday temperatures.*

These discoveries were all made after the war ended. Time and money were available so that material growth experts could work on producing pure samples of silicon. All the other types of atom in the crystal that were not silicon atoms had to be removed first. These are called *impurity* atoms. Then the theoretical predictions, which had first been made in 1931, could be tested. Pauli had been right, to some extent. Semiconductors were a 'filthy mess' because all the crystals studied from Braun's day to 1947 contained too many impurities.

So what did the physicists do as soon as all the impurity atoms had been painstakingly removed from silicon? They put the impurity atoms back, one type of impurity atom at a time. A small amount of one type of solid element was placed on the silicon slice and heated to melting. The random jiggling of the impurity atoms increased and some melted into the silicon. Three discoveries resulted: first, a new way to turn silicon into a conductor, second, a new type of conductor and third, a new, revolutionary electronic device.

What is in a name?

The name impurity does not do justice to the amazing results a small number of atoms of one element can achieve in a semiconductor. When I changed my field of research from high energy physics to solid state physics, I soon realised how much more difficult it was to raise research funding in the new area. It came to me that one reason for this could be that my old colleagues were better at finding more exciting names for their discoveries.

THE HISTORY OF THE SEMICONDUCTOR REVOLUTION

I well remember the first time that this insight struck me – almost literally. It was a cold, wet and windy lunchtime in Kensington Gardens and I was jiggling around on the spot trying to keep warm like an atom in a semiconductor. It was my first game for my new group in the Physics intra-departmental, five-a-side football competition. It suddenly struck me how foolishly inappropriate it was to be shouting 'come on Solid State' to encourage my new team mates, when I had exhorted my old colleagues with 'come on High Energy'.

The importance of a good name had been an early lesson to me as a young PhD student in what was then called high energy nuclear physics; the subsequent dropping of the n-word is but one example. The following, probably apocryphal, story was told to gullible students like me. Two theoretical physicists, Gell-Mann and Zweig, independently proposed that protons and neutrons are made up of the same type of sub-particles but Gell-Mann was awarded the Nobel Prize because he called them 'quarks' while Zweig had proposed the rather more pedestrian name 'aces'. Even if the story is not true, the exotic names chosen for later quark discoveries suggest that other particle physicists believed it: 'Strangeness', 'Charm', 'Truth' and 'Beauty'.

My new subject, solid state physics, had clearly not learnt this lesson. The very name 'semiconductor' suggests a half-baked conductor. A better name may be 'dual-conductor'. The name 'impurity atom' sounds derogatory for an atom which we will see had revolutionary implications. In solid state physics most of the work is done by 'negatively' charged electrons.

When semiconductor physicists discovered a positively charged particle that conducted electricity they decided to call it a 'hole'. This is a quantum particle in many ways equivalent to particle physics' more dynamic 'positron'. A 'Department of Hole Engineering' would be ambiguous at best. A 'Department of Electronic and Positronic Engineering' would surely attract funding.

THE MYSTERY OF THE QUANTUM CONDUCTOR

In semiconductors, electrons and holes 'drift' in an electric field, whereas in high energy physics they 'accelerate'. While being struck by this thought, I was nearly struck by the football. I then had to show some 'high mobility' – solid state physics got that one right!

The benefits of diversity

When the early semiconductor physicists reintroduced small amounts of just one type of impurity atom into pure silicon, the electrical properties were transformed. To help my analogies I will use the more descriptive and more positive name of *diversity* atom.

> *Revolutionary discovery 1*: Adding a small number of some types of diversity atom to an insulating silicon crystal turns that part of the crystal into an electron conductor.

To perform this amazing trick, a diversity atom must have five electrons in the outer orbit – one more than silicon. The atom called phosphorus is one such example. If a phosphorus atom could lose one electron, it would to all intents and purposes act like a silicon atom. It could take the place of one atom in the silicon crystal, sharing its remaining four electrons with the surrounding silicon atoms. It can bond strongly with neighbouring silicon atoms just like a silicon atom. The neighbours would think the new arrival not so bad after all and welcome him or her into Silicon Street. True the newcomer has one more proton in its nucleus than your normal respectable silicon atom, but, as long as they have four electrons to share, the newcomer will be accepted and bond strongly with the neighbours.

Another reason I prefer the name diversity atom is that the citizens of Silicon Street would fear for the reputation of their neighbourhood if a new arrival were known by their scientific name of *dopant*.

How does the new arrival get rid of the embarrassing extra electron? Why, push it into the empty conduction band of course! Now here comes the first lucky break without which there would not have been a semiconductor revolution:

Crucial coincidence 1: Raising the unwanted electron in some diversity atoms up into the silicon conduction band uses about the same amount of energy as an electron gets from the random jiggling of atoms at ambient temperatures.

It is as if the adults directing the street party decided that the 'naughty step' for children who misbehaved on the ground floor was just one step down from the top of the stairs leading to the first-floor playroom. Hence the natural exuberance of the children in 'time-out' on this step might mean they jump into the playroom and turn the community centre from an insulator into a conductor.

This really is one of the most technologically important of a number of coincidences in physics. The idea that the average energy of motion of a jiggling atom only depends on the temperature was a cornerstone of nineteenth-century physics and works well for atoms in solids and in gases. The energy needed to raise an electron from a diversity atom to the silicon conduction band has nothing to do with temperature and depends on Bohr's quantum mechanical ideas. It is a coincidence that these energies are very similar. If they were not, your laptop and mobile phone would not work.

These experiments with phosphorus atoms were exciting. They suggested that a new type of electrical circuit can be formed, a pathway for electrons *inside* a silicon crystal surrounded by insulator, just like a wire but on a much smaller scale. However, the next experimental observation was far more exciting.

The atom boron has three electrons in the outer orbit – one less than silicon. If it could find an extra electron, it would to all intents and purposes act like a silicon atom. In particular, it could replace a silicon atom in the crystal. It could share its four electrons with the surrounding silicon atoms and bond with them. In this case, the neighbours would think the new arrival in Silicon Street slightly odd – one proton short of a respectable silicon nucleus. But as long as the new arrival was happy to share four electrons with the neighbours, he or she would be welcomed into Silicon Street and bond strongly with the neighbours.

Where might a new arrival find an extra electron in a silicon crystal? Do any corner shops in Silicon Street stock them? There are plenty of electrons in the full valence band doing a great job holding the street together. Here comes the second lucky break necessary for the semiconductor revolution:

Crucial coincidence 2: Raising an electron from the valence band to the outer orbit of some types of diversity atom uses about the same amount of energy as an electron gets from the random jiggling of atoms at everyday temperatures.

In this case, the naughty step is just one step *up* from the ground-floor playroom on the stairs leading to the first floor. Any children whose enthusiasm makes them jiggle up and down excessively while playing musical chairs will find themselves in time-out on this step.

Would anyone notice the new arrival in Silicon Street sneaking an electron in order to be acceptable to its neighbours? There are trillions and trillions of electrons in the valence band and only a few diversity atoms are required to produce these exciting effects. The odd electron will not be missed, and there is absolutely no danger Silicon Street will fall apart. However, just one electron missing from a full electron band has a revolutionary effect.

This fascinating part of our story is pure quantum physics. The street party analogy comes into its own. Imagine the start of the game where all the children have seats facing outwards in a large circle. This represents a full valence band holding the street together. One of the children starts misbehaving, so you give them time-out on the boron naughty step. This is one step up from the bottom of the staircase. To maintain order, you insist the game waits for the miscreant to return. While waiting, the children hatch a cunning plan to take revenge on the troublemaker and on you, the organiser.

Just as the misbehaver returns to the empty chair, one of the two children sitting next to the empty chair slips into it. The next child slips into the seat just vacated and so on ... Because one diversity boron atom has removed one electron from a full valence band a very important change has occurred. All the remaining electrons can move ... at last!

> The valence band, while holding the crystal together, can conduct electricity.

But there are more unexpected quantum effects to be noted. Think carefully about this new game. Once they get the idea, all the children co-operate to keep the empty chair moving tantalisingly ahead of the miscreant running round the outside of the circle. These children have worked out a piece of quantum physics vital to the semiconductor revolution:

> The empty seat is moving round the circle as fast as all the children are moving but *in the opposite direction.*

If you apply an electric field to a full valence band with one electron missing all the electrons can move one way and hence they form an electric current. The valence band – the band that is holding the crystal together – has turned into a band that will conduct electricity just because one electron has been removed. It is a completely new type of electricity resulting from the absence of just one electron. The vacancy is moving in the opposite direction to the negatively charged electrons. It is acting like a *positively charged particle.* Solid state physicists call the new particle, a 'hole'. I prefer the particle physicists' name for a quantum particle formed in this way – a 'positron'.

> *Revolutionary discovery 2*: Adding a small number of some types of diversity atom can turn a region of silicon from insulator into a new type of conductor, a positron conductor.

These two revolutionary discoveries resulted from readmitting a small number of two types of diversity atom into silicon. Different parts of the crystal can be turned into either electron conductors or positron conductors just by adding a small number of diversity atoms in each part. It only takes a very few newcomers to change the character of the street. With modern crystal growth techniques, insulators can be turned into electron conductors or into positron conductors from one atomic layer to the next atomic layer. This can be done by adding one diversity atom to every thousand silicon atoms, or one diversity atom to every billion silicon atoms or all values in between.

A pure semiconductor, which is an insulator, can be turned into an old-style electron conductor in one region and into a new-fangled positron conductor in another region. This turns out to make many exciting new electrical devices possible. I think semiconductors should be more appropriately, and more attractively, described as *dual-conductors*.

The semiconductor diode

Two revolutionary discoveries and two crucial coincidences resulted in it being possible to produce regions of electron conductor and regions of positron conductor in the same semiconductor crystal. A third discovery was even more exciting and it came when researchers studied the region between an electron conductor and a positron conductor:

Revolutionary discovery 3: The region between an electron conductor and a positron conductor acts like a diode valve. In a diode, electric current can only flow one way.

Let's see how this works. Assume a small number of diversity atoms have turned one side of Silicon Street into an electron conductor and the other side into a positron conductor. Life in Silicon Street gets a lot more interesting. The electrons introduced into the conduction band on one side of the street are free to move up and down the street and can perform some useful work. On the other side of the street, the new arrivals have come from a very different family tradition. In their case, it is the conducting positrons that go out to work. However, coming from the valence band, they never neglect their responsibilities to keep the home together.

While fitting in with the social norms of Silicon Street, the electrons and the positrons remain aware of their cultural heritages. The electrons always remember the price of making it to Silicon Street. The nucleus back home has one more proton than is the norm on Silicon Street. The home of each of the new arrivals has a positive charge. Similarly the positrons on the other side of the street never forget that when they moved into Silicon Street their nuclear family was one proton short of a respectable silicon atom. This is indelibly etched into their subconscious. The homes of new arrivals on their side of the street are negatively charged.

With the diversity homes having positive charges on one side of the street and the diversity homes on the other side having negative charges, there will be an electric field permanently acting from one side of the street to the other.

This electric field makes the electrons go one way across the street and the positrons go in the opposite direction. The flow of electrons one way and positrons the other way makes up an electric current. Remember musical chairs; the empty chair moves in the opposite direction to the electrons.

A device that makes a current flow one way is called a diode or a rectifier. Valves that operate in this way were vital to the development of wireless as we saw earlier and to all areas of the old-fashioned electronics.

This third discovery was truly revolutionary. Because the two types of conductor could be manufactured close together in the same crystal, it became possible to make a diode valve *inside* a silicon crystal: no fragile glass vacuum tube, no glowing heater to drive the electrons out of one contact, no bulky power supply to provide the field or heat the electrons. All these functions could be manufactured *inside* the silicon crystal by the controlled addition of two types of diversity atom. The diversity diode is the basis of all devices in the semiconductor revolution.

In the next chapter, we will find that two diversity diodes formed back to back but facing in opposite directions worked as a new device, a *transistor*. It performs the functions of a triode valve and so can amplify radio signals.

We will also see how eventually a way was found to manufacture whole electronic circuits with many transistors in one small crystal. Once that breakthrough had been made, the devices and the circuits could get smaller and smaller until they reached atomic sizes, a situation the technology is now approaching in the early twenty-first century.

All this was possible because Feynman's 'remarkable mystery' had been solved. Electrons (and positrons) move nearly as freely through a semiconductor crystal as in a vacuum thanks to the electrons being both particles and waves, jumping and resonating from atom to atom. This explanation of the mystery led to the realisation that diversity atoms could turn an insulator like pure silicon into an electron conductor or a positron conductor in a controlled way.

THE HISTORY OF THE SEMICONDUCTOR REVOLUTION

We will meet one more revolutionary discovery and two more crucial coincidences in the next chapter. These, with help from Einstein's equation $E = hf$, give us the solar cell. In its simplest form, the solar cell consists of just one diversity diode. We have solved one mystery but that other nagging question remains: why have solar cells been exploited so much more slowly than transistors?

SIX

The Semiconductor Revolution

> The transistor and the development of solid-state electronics are technological offspring of the revolution in physics which began with Planck and Einstein at the beginning of this century. I am proud to be able to say that it has all happened in my lifetime.
>
> Walter Brattain

Whatever your age, your life will have been affected by the semiconductor revolution. You might recall the sensation caused by the first transistor radios in the 1950s. Today, social network sites have transformed many of our friendships, while, for many, computer games dominate entertainment. You might be reading these words in an electronic book or downloaded from the web. None of this technology would have been possible had not quantum physicists solved the mystery of the two ways that semiconductor crystals conduct electricity.

The impact of semiconductors can truly be described as revolutionary. The mobile phone has been a major factor in popular revolutions in North Africa and the Middle East. A top-selling newspaper in Britain has been closed because of allegations that its reporters abused mobile phone technology. These words are written, not with pen and ink, but hunched over a laptop.

Semiconductors made computers much smaller and faster. As we will see, $E = hf$ also made it possible to convey information between computers by using light in fibres instead of electrons in wires. Quantum ideas also helped wireless technology shrink from Cornish cliff-tops into devices we can fit in our pockets.

THE HISTORY OF THE SEMICONDUCTOR REVOLUTION

The diversity diode we met in the last chapter is the basis of all the devices in the semiconductor revolution. The revolution started with the invention of a *transistor* consisting of two such diodes back to back. Within 10 years of their discovery, transistors started replacing valves, which were eventually consigned to science museums. Then, even more amazingly, whole circuits containing many transistors were manufactured *inside* the crystal. Circuits got smaller and the number of transistors got larger until in 2010 one flat, silicon crystal contained a computer consisting of a billion transistors.

Smaller circuits do not just mean more calculations and more storage of information in a crystal. They also meant faster calculations and faster release of information because electrons and positrons don't have to move so far. I get annoyed when modern search engines come up with hundreds of examples of the data I am searching for before I have finished typing the question. I should rather wonder that this has all been made possible because quantum ideas have made transistors so small.

We will find that the diversity diode is also the basis of the solar cell, which we will meet in this chapter. It is one of the simplest of semiconductor devices and also the most direct form of electricity generation.

Take-off for the semiconductor revolution

The first working semiconductor device suitable for mass production was the transistor invented by John Bardeen, Walter Brattain and William Shockley at Bell Laboratories in New Jersey in 1947. All three were awarded the 1956 Nobel Prize in Physics. The theorist John Bardeen is the only person to have won two physics Nobel Prizes.

The transistor consisted of two diversity diodes back to back like the slices of bread in a sandwich. Crucially, the two diversity diodes faced in opposing directions. It operated like a triode valve. A varying voltage, for example a radio signal, applied to the filling in the sandwich, can be amplified. The enhanced signal appears in the current passing between the top slice and the bottom slice of bread.

I can dimly recall the sensation the new transistor radios caused in the late 1950s. The old-fashioned valve radios were large and heavy; the impact of a small, cheap portable radio was exciting, particularly for young people. This was a relatively slow take-up compared to some of the achievements of the semiconductor industry in more recent years. John Orton, my boss in the year I spent at Philips Research Laboratories, has written a fascinating history, *The Story of Semiconductors*. He points out that the sale of old-fashioned valves peaked in 1957 – ten years after the transistor was invented.

Orton also notes that in those early years the semiconductor industry was kept afloat by the US military. Cost was not an issue for them. Their missiles needed electronics that were light and reliable rather than cheap. In 1961 two US companies made a breakthrough in providing smaller, lighter electronics: the *integrated circuit*. Circuits were laid out on the top surface of the silicon wafer. This approach was easier to manufacture reliably than the transistor and all the wire contacts were on the top surface. Most importantly, the approach had the potential for further miniaturisation. But, as Orton points out, the early integrated circuits were very expensive.

The new industry might have floundered before reaching commercial viability had it not been for US strategic ambitions, in particular, President John F. Kennedy's decision in May 1961 to put a man on the moon by the end of the decade. As Orton puts it 'such a dramatic kick-start to a technological revolution smacked of divine intervention by a Higher Being ...'

A new type of transistor was particularly suited to the new integrated circuits. This was called the *field-effect transistor*. It worked just with electrons, directly mimicking de Forest's triode valve and operating near the surface of the crystal. An equivalent field-effect transistor was developed that only used positrons.

A step-change in the semiconductor revolution occurred in the late 1960s. This was a result of having separate transistors made from the traditional and the new form of electricity. A new family of devices known by the acronym CMOS (Complementary Metal Oxide Semiconductor) was developed. This combined an electron transistor and a positron transistor in a clever way. For the first time, information could be stored electronically (and positronically!) with electrical power only being necessary when information is put in, or taken out of, the store.

If you are reading this on an e-book or downloaded on a laptop you can reflect on the fact that these words can only assemble on the screen if trillions upon trillions of positrons as well as electrons have done their job properly. Thanks to CMOS, equality between positron conductors and electron conductors began in late 1960s' semiconductors.

We all have our own stories about the ways the semiconductor industry has changed our lives, and how rapidly the technology changes in a generation. My story is that when I started my PhD at Birmingham University in 1964 I chose experimental particle physics not just because the physics was exciting and there was a chance of working at CERN in Geneva, but also because there would be the opportunity to learn how to use the latest computer technology. In fact, the Birmingham University particle physics group did not have its own computer at that time. They were too expensive. We were given access to one of the state-of-the art computers at Glasgow University. Once a fortnight we flew from Birmingham to Glasgow clutching boxes of cards containing our programs and stayed up all night to run them. That single computer was impressive, with its many cabinets filling a large room. Twenty years later in 1985, when I spent a year in Philips Research Laboratories, every member of the technical staff had a personal computer on their desk, each with more memory and faster computational speed than the room in Glasgow. Twenty years later in 2005 every politician and top executive had a mobile phone with even more memory that could be accessed much faster.

To infinity and beyond!

We have so far considered the part of the semiconductor revolution concerned with computers. Many other applications are possible because there is one more extremely important way that a semiconductor can be turned from insulator to conductor. In this case, it turns into both an electron conductor and a positron conductor.

Back to the Silicon Street children's party in the community centre. Remember that the musical chairs game on the ground floor and the playroom on the first floor represent the valence band and conduction band respectively in a silicon crystal.

Why not liven up the proceedings? Prearrange for a friend dressed as Buzz Lightyear to arrive when the musical chairs game starts. Buzz could make a dramatic entrance through the window on the ground floor of the community centre. His job is to pick up the most bored child from musical chairs and carry him or her on his back up to the first-floor playroom where the child can run around and let off steam. Note also that the remaining children on the ground floor are now free to move around from chair to chair. Just one child carried up to the first-floor playroom transforms the Silicon Street community centre.

One electron raised from the valence band to the conduction band will transform a semiconductor into an electron conductor and a positron conductor.

Let's think about what helpful tasks the lucky child might perform on the first floor of our transformed community centre. Of course, they would have to be under the direction of an adult, who represents an electric field. The lucky child could accompany a helper when they go out of the door on the first floor of the community centre to the corner shop to buy some more balloons. The child then completes the circuit by returning, with his or her supervisor, through the entrance on the ground floor to the musical chairs game. The adult directing the musical chairs will have arranged that all the seated children shuffle round from chair to chair so that the empty chair moves to the point nearest the ground-floor entrance. The returning child can then flop down into the empty seat. The electrical circuit has been completed on the ground floor by the movement of the empty chair or positron. The Silicon Street community centre, and the silicon crystal, return to their original states.

If we could find an equivalent to Buzz Lightyear to provide the energy to raise electrons from the valence band to the conduction band and an internal electric field to do the job of adults directing the children, we would have created an exciting new device. It would be a semiconductor crystal that could drive electrons round an external circuit and do useful electrical work.

Creating an internal electric field is easy; that can be achieved by having different diversity atoms on the two sides of Silicon Street. Finding enough Buzz Lightyears is even easier. Buzz Lightyear represents one of the super-heroes of this story – a photon of sunlight. Remember, enough photons of sunlight fall on the earth in an hour to supply all of humankind's energy needs for a year: Lightyear by name, light year by nature.

The popular name for the silicon crystal I have described is a *solar cell*.

Revolutionary discovery 4: Just one photon of sunlight absorbed by a semiconductor containing a diversity diode can transform the semiconductor into a solar cell.

THE SEMICONDUCTOR REVOLUTION

The electron and the positron form an electric *current* because the electric field of the diversity diode pulls or pushes them in opposite directions. The energy in the photon has raised the electron to a higher energy of position than it had in the valence band. This means there is an electric *voltage* between the conduction band and the valence band, which produces another electric field which can drive electrons round an external circuit. This is why the technical name for a solar cell is a *photovoltaic* (PV) cell. A solar cell produces both current and voltage; a solar cell directly produces electric *power* (which you may recall is current *times* voltage).

The photon has to have energy equal to, or higher than, the band-gap energy of the solar cell. Superhero he may be, but Buzz has to have enough energy to carry the child up the staircase. Now here comes the first of two amazing coincidences about the band-gap energy of silicon, the energy difference between the ground-floor and first-floor playrooms of the Silicon Street community centre. The red photons of sunlight, at the lower energy side of the rainbow, have higher energy than the band-gap energy of silicon. According to the equation $E = hf$, violet photons have even higher energy – nearly twice the energy of red photons.

Crucial coincidence 3: Every photon in sunlight which our eyes can see (red to violet) has sufficient energy to raise an electron from the valence band to the conduction band of silicon.

Remember that the sun is golden because the gold photons are most numerous in sunlight. Gold photons have energy between red and blue photons. Remember also that Planck used the equation $E = hf$ to show that the temperature of the sun fixes the number of photons of each colour in sunlight.

Crucial coincidence 4: The temperature of our golden sun is such that the band-gap energy of silicon, one of the most abundant elements on earth, is nearly ideal for converting the energy in sunlight to electricity efficiently.

These coincidences are worthy of John Orton's description of the development of integrated circuits; they smack of 'divine intervention by a Higher Being'. If life on earth is to continue, we need to learn from these lucky coincidences revealed by the equation $E = hf$.

So why didn't the solar revolution start in 1947 along with the semiconductor revolution? In fact, early researchers were well aware of the potential for electricity generation in a solar cell made possible by these discoveries. Indeed in 1954 Bell Laboratory researchers Daryl Chapin, Calvin Fuller and Gerald Pearson produced the first practical PV cell. Their silicon cell converted 6 per cent of the power in sunlight to electrical power. This was extremely promising for a first attempt. As we will see, solar cells did not take off anywhere near as quickly as did transistors. Sadly, the inventors have yet to receive a Nobel Physics Prize for their discovery.

The observation that a single photon could turn an insulating silicon crystal into an electron and positron conductor did have one early commercial application. If an electric field is applied to a solar cell so that it enhances the internal field of the diversity diode, you have a *photodetector* or 'electric eye' as it was known in the 1950s. If a beam of light shines on the electric eye, a current will be generated. If a person walks through the beam, the current drops and clever electronics can ensure a door opens or closes automatically.

For those who cannot remember the sensation caused by the first electric eyes, I recommend my favourite film, which dates from that period: Jacques Tati's *Mon Oncle*. Tati's character, Mr Hulot, is bemused by the fashionable gadgetry in his sister and brother-in-law's new home. It includes a garage door activated by an 'electric eye'. The door opens as the couple drive into the garage. But before they can make their exit, a dog with its tail up walks through the beam so the door shuts on them. One of the most hilarious parts of a very funny film shows the couple gesticulating through two porthole windows, trying to get the dog to raise its tail and trot back through the beam.

An analogy for the power of a solar cell

Here is an analogy that shows the elegance and simplicity of the way a solar cell powers an electric circuit.

Hydropower is already a major contributor to the solar revolution, as we will see in Part II. Yes, hydropower is a solar technology. Trillions upon trillions of photons of sunlight fall on the sea each second increasing its heat energy. The water molecules jiggle around randomly more and more as the temperature of the sea rises until some water molecules have enough energy of motion to leave the ocean and form water vapour. As we will see in Part II, the hot water warms the air, which expands and rises. Hence water molecules are carried up into the sky where they form clouds of tiny water droplets.

The hotter the sea becomes, the more water molecules there will be in the air. Global warming, which we will also discuss in Part II, is therefore leading to wetter winters.

Winds blow the clouds over the sea to the land. Here the energy of position that water molecules have because they are in clouds high above the ground is converted into the energy of motion of raindrops as they fall on the hills. The water runs down hill and builds up behind a dam. The energy of position of the water behind the dam is converted into the energy of rotation of a turbine, which can generate electricity using Faraday's ideas. Finally, the water molecules end up in a river flowing back to the sea where the cycle starts again. Hydropower gives us effective and efficient renewable power when it is needed from the solar energy that falls on the sea.

THE HISTORY OF THE SEMICONDUCTOR REVOLUTION

A solar cell circuit is a much simpler way of generating electrical power from sunlight, but there are similarities to hydropower. One photon of sunlight will raise one electron to the conduction band leaving a positron behind in the valence band. This is a much more straightforward process than a water molecule being carried up into the clouds. Much of the energy of the photon is directly converted into the energy of position that the electron has in the conduction band compared to the energy of position it previously had in the valence band. A diversity diode will produce an *internal* electric field, inside the solar cell, which makes the electron and positron move in opposite directions and forms an electric current.

The equivalent of the wind blowing the clouds to the land is the *external* electric field in the circuit. A physicist says this is because the electrons have higher energy of position in the conduction band than the valence band. An electrical engineer says there is a voltage difference between conduction and valence bands. On either picture, there is an electric field in the external circuit, which forces electrons to move through the wire much more directly than the wind forces clouds to move.

The electrons from the conduction band have higher energy of position just like the water molecules in the clouds. This can be converted into the energy of motion of electrons in an electric motor and hence to the rotational energy. This is like the energy of position of water molecules behind a dam being converted into the rotational energy of a turbine. Thanks to the electric field in the return wire, the electrons flow back to the valence band of the solar cell just like water flows from the turbine back to the ocean.

The solar cell circuit is completed by a positron moving through the valence band of the cell to the return wire under the influence of the electric field of the diversity diode. Here an electron from the wire drops into the vacancy. The electron and the positron are unlikely to be the original ones as they both move slowly compared to the electric field. That does not matter. The circuit is completed and the solar cell returns to its original condition.

Having explained this analogy, it is important to emphasise that, in addition to its simplicity and elegance, photovoltaic power has an additional advantage over hydropower and most other forms of renewable energy. A solar cell can generate current and voltage *at the place the electrical power is needed*. The technical name for this is *distributed generation*. Compare the physics of the solar cell described in this chapter with the description in the last chapter of the way the electric field is transmitted hundreds of kilometres to boil a kettle.

Here is a third possible solution to Hawking's question in the Introduction. Extant alien civilisations, which found $E = hf$ first, may be observing us from afar. They note that influential earth scientists, politicians and even some environmentalists favour electricity generation from $E = mc^2$ hundreds of kilometres away in a nuclear reactor rather than local generation of the electricity from sunlight. They have decided earthlings are too stupid to be worth colonising.

You cannot make bombs with solar cells

Solar cells are simple, low carbon, electricity-generating devices. They are produced by the industry that has revolutionised our lives. Most solar cells are made from an abundant material, which has near ideal band-gap energy for producing electricity from sunlight. What is not to like? How come we aren't all using them to generate most of our electricity? We have encountered clues to the answer in earlier chapters and we will uncover more, but perhaps the simplest answer is that you cannot make bombs with solar cells.

Many of the semiconductor gadgets that we now take for granted were developed first for the military and hence took off faster than solar cells. Integrated circuits were one example I discussed earlier. The digital camera is another. If the specially prepared surface of a silicon crystal is covered by a large number of small diversity diodes, any photons arriving at the surface can produce electrons, which can be stored in the diodes.

The image of a distant object is focused by a lens onto this surface. The number of electrons in any one diode will depend on the number of photons hitting that diode. Diodes in the light part of the image will contain many more electrons than the diodes in the dark part. Hence a digital representation of the image is formed. Clever electronics can work out how many electrons there are in each diode. This is called a charged coupled device (CCD). It is the basis of the digital camera.

The CCD was invented at Bell Laboratories by Willard Boyle and George Smith in 1969. They were eventually awarded the Nobel Physics Prize in 2009. The first practical application came in 1976 when CCDs were used in a US spy satellite to produce real-time surveillance images. Many governments, in addition to the US, have been happy to pay the large development costs as the military found them particularly useful for night vision.

It was some time before public demand for digital cameras took off and mass production beat down the price. In 1994 I bought a system to measure solar cells, which contained a CCD costing around £2,500 to manufacture. At that time, CCDs were also used in television cameras. Seven years later, in 2001, I bought my first digital camera for about £250. The active part of this camera was a more sophisticated CCD than the one that cost ten times more seven years earlier. Seven years on, in 2008, I was offered a free mobile phone by a wireless provider if I took out a new contract. I asked if I could have one *without* a digital camera. I was told no. A much more sophisticated version of a CCD system, which 14 years earlier had cost £2,500 to make, was being given away. This shows what the semiconductor industry can achieve in 14 years given domestic demand and mass production, providing the development costs have been paid by the military.

The technology of silicon solar cells and CCDs have many similarities. They both absorb photons of sunlight in diversity diodes.

Solar cells have the advantages over CCDs in that they don't need clever electronics to work out how many electrons there are in each diode. On the other hand, CCDs have the advantage that they cover a very much smaller area than solar cells. The smaller the device, the easier it is to mass-produce and the cheaper it becomes. Silicon solar cells are simpler technology and are straightforward to mass-produce but they have to cover a large area so the amount of silicon is an important factor in the cost.

Though you cannot make bombs with solar cells, the military did realise that they could be useful in other ways. There is plenty of sunlight in space and solar cells can provide the power that a military space satellite needs to run its CCD cameras. The US military, NASA and the space programmes of other nations funded much of the early development costs of solar cells. In the case of integrated circuits, which I discussed earlier, NASA and the military wanted smaller, lighter systems, which turned out to be suitable for mass production. In the case of solar cells for space, cost was not the important factor; cell efficiency, light weight and reliability were what mattered. Solar cells for domestic use have not benefitted as much from the space programme as did integrated circuits. The cost of rooftop PV did not come down until the new solar cell industry faced up to the one big, technical problem that threatened the fledgling solar revolution.

The indirect band-gap problem

Not all the problems facing the young PV industry were political. There was a major technical problem with silicon, which the young solar cell industry had to solve. It will help understand the challenge if we pay another visit to the Silicon Street community centre, which represents a silicon crystal.

If it was 'divine intervention' that the height of ceiling in the Silicon Street community centre is ideal for a solar cell, the Creator could have done a better job with her architectural plans. The first-floor playroom in the community centre may be at the ideal height, but it is not in the ideal position. It is not directly above the ground-floor playroom, but rather on the other side of the building. Silicon is what is known as an *indirect band-gap* semiconductor. This means that, after carrying a child upstairs on his back to the first floor, Buzz Lightyear has to stop and draw breath, cross the hallway and carry the child through into the playroom. No big deal, except that it means silicon does not absorb sunlight particularly well. A crystalline silicon solar cell has to be about half a millimetre thick to absorb all the sunlight. That doesn't sound thick and it isn't. The problem is the silicon has to be high quality, which means it is expensive. The thicker the crystal, the higher is the cost.

In the early years of the semiconductor revolution high-quality silicon crystals were assumed to be necessary for good solar cell performance. Both the electrons and the newly discovered positrons have to hang around for some time to be given directions by the electric field. Poor quality crystals contain what are appropriately called *traps* where unsuspecting electrons and positrons may be detained indefinitely while awaiting directions from the field. To get rid of the traps the crystal had to be purified and that takes energy.

Some early crystalline silicon solar cells spent a considerable proportion of their working life generating electricity to repay the energy used in purifying the crystal. That didn't matter for cells used on satellites as the cost was not the important factor. This criticism is still heard from opponents of solar electricity, despite the solar cell energy repayment time having fallen dramatically. For the latest cells in the domestic market, which have a guaranteed working life of 25 years, the energy used in their manufacture is repaid within 2–4 years. Bear in mind also that after fossil and nuclear fuel generators have paid back the energy used in their manufacture they still need energy input; they burn fuel.

Solving the indirect band-gap problem

While her older sisters, dressed in glittering crystalline silicon, were having a heavenly time at the satellite ball, poor Cinderella, the domestic solar cell industry, was left behind on earth to make the best she could of two cheaper types of silicon. One of them was a hand-me-down from rich cousins in the silicon chip business.

One of her two options, *amorphous silicon,* has no crystal structure at all. The material was first developed by another German émigré, Walter Spear at the University of Dundee in Scotland. It does have a band-gap and, very luckily for PV, it is direct; the playroom is above the circle of chairs in the community centre. That means very much less than a hair's breadth, and even less than the thickness of household cling-film, will absorb all sunlight.

Amorphous silicon cells were the first example of *thin film* solar cells. They are cheaper as they use less material. The downside is that the material isn't as good quality as crystalline silicon. Amorphous silicon cells convert the power in sunlight to electrical power with about two-thirds of the efficiency of crystalline silicon and the cells do not have the latter's 25-year lifetime.

Despite this unprepossessing performance, amorphous silicon gave the solar cell industry its first place in the domestic sun in the late 1970s. When handheld electronic calculators first appeared in classrooms and supermarket checkouts in the 1970s, their batteries needed chargers like our present-day laptops and mobile phones. Then in 1978 versions powered by amorphous silicon solar cells appeared and the chargers were no longer needed. I marvel that my solar-powered electronic calculator has outlasted at least four laptop computers and three mobile phones. It is powered by the ambient light on a small area of solar cell, which is covered up for lengthy periods.

The second cheaper version of silicon, Cinderella's hand-me-down, is called *polycrystalline silicon*. Again the name is appropriate as a single crystal actually consists of many smaller crystals. Clever chemical processing of the crystal ensures that the electron and positron traps between the sub-crystals are less important. Respectable power efficiencies (around 14 per cent) are typical in these cells. Being of lower quality and requiring less energy input, polycrystalline silicon is cheaper than crystalline silicon. It is also cheaper because, at least until recently, the young PV industry used silicon produced for, but discarded by, the silicon chip industry.

Take-off for the solar revolution

When, at last, towards the end of the twentieth century, the governments of Germany and Japan decided to kick-start the solar revolution with incentive schemes, polycrystalline silicon solar cells took their place in the van of the revolution. Demand grew, mass production took over and the price of the cells started falling as these governments had expected. Then, in 2004, costs stopped falling as a polycrystalline silicon shortage developed. This was because the expanding solar cell industry was at the point where it had raised the demand for polycrystalline silicon to the rate at which the silicon chip industry was discarding it.

Anticipating this blockage, new solar cell companies had been tooling up to join the market with second-generation thin film cells made of different semiconductors. The new entrants were *cadmium telluride* (CdTe) and *copper-indium diselenide* (CIS). Like amorphous silicon, both CdTe and CIS are direct band-gap, with conduction band playroom above the ground-floor valence band. That means considerably less than a hair's breadth of material is required to absorb all the sunlight. Potentially they are much cheaper than the first-generation polycrystalline silicon cells. A US company, First Solar, led the way with CdTe cells manufactured in the US and Malaysia. In Part II we will see that they were the first company to manufacture cells totalling one gigawatt (GW) of power in one year.

At last, in early 2009, the price of PV cells started falling again. It continued falling in a dramatic fashion as new, dedicated factories manufacturing polycrystalline silicon came on stream. Another reason for the decline in price was the arrival from left-field of a number of Chinese companies with 1 GW a year production lines for first-generation cells. We will consider the dramatic effects of these developments in Part II.

Now that we are up to date, let's return to the 1960s to see how another part of the semiconductor revolution solved the indirect band-gap problem. Their solution turned out to be important for third-generation solar cells and for mobile phones.

Go directly to 'Go'

There are many semiconductor devices other than solar cells that need to be good absorbers of photons and also need to be cheap. Any device that absorbs photons well will emit photons well. So semiconductor devices that emit photons should also have a direct band-gap.

A whole new industry, known as *optoelectronics*, grew up in the 1960s to develop and manufacture photon absorbing and photon emitting devices. The new industry was based on another, more expensive, semiconductor known as *gallium arsenide* (GaAs), which has a direct band-gap.

GaAs crystals are more expensive than silicon ones because they are grown using more sophisticated methods. Cost has not been a big problem for the optoelectronics industry because the devices are usually very small and do not need much of the expensive material. Also, as we will see, the military, and therefore governments, are very interested in some optoelectronic devices.

Photon emitting semiconductor devices work like a solar cell running backwards. Recall that a solar cell is simply a semiconductor with a suitable band-gap to absorb sunlight and a diversity diode that forces electrons and positrons to move in opposite directions. If an electrical power supply is connected to a solar cell so that the external voltage *opposes* the diversity diode voltage, the electric field of the diversity diode can be reversed. This will force the electrons in the conduction band and positrons in the valence band to change direction. Instead of moving away from each other, they head towards each other. If an electron sees an empty space down in the valence band it can lose energy by falling into it.

Remember that the absorption of a photon in a semiconductor leaves a positron in the valence band and an electron in the conduction band. The reverse also occurs. An electron in the conduction band can drop down into the vacancy that formed the positron in the valence band. The energy lost by the electron is emitted as a photon. In the language of particle physics, the electron and positron have annihilated to form a photon.

Imagine the Silicon Street community centre solar cell working backwards. The adults upstairs direct a child to jump onto Buzz's back. He or she is then carried downstairs and dropped back into the empty chair. Buzz uses the energy gained in running downstairs to fly out of the ground-floor window.

But don't marvel too long at Buzz's superhero powers ... this carefully organised street party is turning into a disaster! Hoards of uninvited children are arriving through the door on the first floor thanks to the external voltage. Even worse, many of the children playing musical chairs are now rushing out the ground-floor exit leaving empty chairs behind. This is all because the external electric power has spoilt the well-laid plans of the organising team. It has pointed the children in the opposite direction. My analogy for this quantum process is starting to creak. We need many Buzz Lightyears waiting upstairs to carry the new arrivals down to empty chairs and then all buzz off out of the ground-floor exit.

The device I am describing is a *light emitting diode* (or LED), the reverse of a solar cell. The extra energy of position that electrons have in the conduction band compared to the energy that they have in the valence band, is converted into the energy of a photon of light. The electrons lose energy by dropping down to fill an empty chair in the valence band. In the case of GaAs crystals, the equation $E = hf$ says the photons are infrared ones: just below the energy of red light.

A lamp that emits in the infrared is useful for the military, for the security industry and cats but not much use for human beings. Fortunately, John Orton's Higher Being was smiling on the fledgling optoelectronic industry in the 1960s as she did on the integrated circuit industry in the same decade. Materials scientists soon came up with pure crystals of a whole new family of direct band-gap semiconductors. They were made from compounds of the elements gallium and arsenic with other elements that had the same number of electrons in their outer orbits. These new *alloys* of GaAs as they were called bond into strong tetrahedral crystals just like silicon. New production techniques were developed in which the chemical composition and band-gap of the crystal could be changed at will. Within three decades, LEDs would able to emit light at all wavelengths that the human eye can see.

LEDs made from an alloy of GaAs were first used as on-and-off lights and on digital clock displays in the 1960s. These early digital clocks had distinctive red numbers. Unlike the old filament or fluorescent light bulbs, they can be turned on and off essentially as fast as the electrical supply.

Applications of LEDs have diversified over the years. Their reliability and speed of response makes them useful in stop lights and signalling lights of cars and in traffic lights. One modern application you may not be aware of is that many of the flat-panel TV screens of the LCD type use LEDs as the primary source of illumination.

Should we be aiming to replace our household bulbs with LEDs? They use a lot less electrical power for the same illumination than the old-fashioned glass bulb, as you can understand from an everyday example. The filament in the old, glass lamp bulbs is a much thinner version of the element in an electric kettle. The electrons in the current passing through the filament bash the atoms in the filament so they jiggle around faster. The temperature of the filament rises much higher than the element in a kettle. Soon the wire glows with visible light like the opening in the cavities studied by Victorian physicists. A lot of the electron energy goes in heating up the filament.

Compare that process with an electron dropping from the conduction band to the valence band and releasing all the extra energy it had in the conduction band as a photon. Because it doesn't have to heat up, an LED is much more efficient than a filament bulb and lasts much longer. Also the emission of photons from an LED can be easily controlled by the applied voltage. In fact the photons emitted can carry information, though that is easier in the laser which we will meet later.

The big breakthough in the use of LEDs for domestic illumination came in the 1990s when Shuji Nakamura and his group at the Nichia Chemical Company in Japan produced LEDs made from gallium nitride rather than GaAs. The new LEDs emit blue light. With blue LEDs, lamp bulbs could be developed which emitted over all wavelengths of sunlight.

I have in front of me an LED light bulb, which I brought in a local supermarket in 2013 for around £5. At first sight, it looks like a halogen spotlight bulb and it produces similar illumination. A number of features distinguish it from the halogen: there are nine small separate LEDs, it has a power requirement of around 1 W rather than 35 W, an anticipated life of 30 years and a price which is currently 10 times higher than the halogen. Later we will find a number of reasons why this price should fall.

Another LED application that is taking off is in torches. The lower power requirements of the LEDs mean torches can run much longer than a conventional filament-bulb torch with the same battery.

The LEDs in the spotlights and torches tend to provide a rather cold light. However, new warmer versions are coming on the market. In late 2012, I attended a presentation about LED lighting given by Philips, the company I had visited over 20 years ago. It was very impressive how versatile the LED is becoming. It is easy to control, not only to turn on and turn off but also in the colour and intensity of the light. All these features should increase demand, thus reducing the price and the carbon footprint of our homes.

Lasers, quantum wells and high mobility electrons

The domestic use of LEDs is an example of a semiconductor technology that has taken off slowly not only because of the indifference of the military but also because a cheap, well-established but much less efficient technology dominates the market.

In the 1960s the military, and many physicists, were much more interested in a more sophisticated version of the LED: the *laser diode*. The laser diode is like the LED, but the surfaces of the crystal are made into mirrors. These reflect photons back into the crystal. The photons bounce back and forth inside the crystal and stimulate more electrons to drop from conduction band to valence band and emit photons. One of the mirrors is designed to let a small proportion of the photons out. A stream of photons emerges in a very narrow beam. All the photons have the same wavelength and all the waves are in step.

I cannot resist pushing the Buzz Lightyear analogy one final time. I promise this is the very last occasion I go 'to infinity and beyond'. The quantum phenomenon of stimulated emission of photons in a laser is like that marvellously surreal moment in *Toy Story 3* when Buzz finds himself in warehouse stacked full of rows of identical, boxed Buzz Lightyear toys.

The first semiconductor laser working at ambient temperature was produced in 1969. The precision of a spot of laser light is ideal for targeting smart weapons, so the military were very interested. The public became aware of lasers in 1979 with the advent of the compact disc (CD) player. This has a laser in the CD reading head.

Lasers have many beneficial medical applications. Thanks to lasers, much of the information shuffled between computers is carried by photons in fibres. Your landline phone messages and broadband are likely to have travelled this way for part of their journey.

New ways to grow GaAs crystals were developed, which led to increased sophistication in laser operation. GaAs crystals can now be grown one atomic layer at a time with the type of atom being changed from atomic layer to atomic layer. Different diversity atoms really can be introduced, one atomic separation apart, on opposite sides of Gallium Arsenide Street.

Even more importantly, in some special cases, all the atoms on one side of the street can be the same, but all of them different from all the atoms on the other side of Gallium Arsenide Street. The number of steps from ground floor to first floor in all the houses on one side of the street can be different from the number of steps in all the houses on the other side. This means that the all-important band-gap can be changed after a few atomic separations. This makes it possible to construct ultra-thin structures called *quantum wells*. They are regions of the crystal only a few atoms wide that have a lower energy band-gap than the regions on either side. The quantum wells were the first artificial *nanostructures* to be successfully employed in semiconductor devices. They are called *quantum* wells because electrons show their schizophrenic behaviour in such narrow confines.

The great advantage of quantum wells for lasers is that electrons and positrons like to fall into a well where there is more chance they will annihilate each other and produce photons. Nowadays no self-respecting laser is without a few quantum wells.

Another GaAs chip that uses just one quantum well makes

our mobile phones work. A quantum well is only narrow in one dimension in the crystal. In the other two directions, the wells can be very wide. It turns out that in the presence of an electric field, electrons move a lot faster along the well than they do in the crystal. The water in a slow-flowing river speeds up as it enters a narrow gorge and slows down when the river widens again.

During my year at Philips I was fortunate to join a group, which for a time held a world record thanks to their excellent quality quantum wells. The team grew such good quality material that the electrons moved faster along the quantum well in a given electric field than achieved by any other of the many groups in the world working on this technology at that time.

This experience was also a warning of the vicissitudes of commercial semiconductor life. A few years later, Philips stopped their world-leading GaAs quantum well research. As a result, the company missed out on one of the biggest booms of the 1990s. Other laboratories, in particular Fujitsu the Japanese company, commercialised a field-effect transistor in which electrons move fast along a quantum well. The device was called, not very originally, the high electron mobility transistor (HEMT).

In one version of the HEMT, the quantum well was formed from a GaAs alloy with a lower band-gap than GaAs and a bigger atomic spacing. In this case, the electrons turned out to move so fast the HEMT became the first transistor to achieve the holy grail of wireless communication foreseen in Chapter 2. The electrons moved so fast along the well, and the HEMT was so small, that it could *directly* amplify the vibrations of a long wavelength electromagnetic wave. Now, at last, Marconi's large aerial could shrink into a device which would fit in a pocket. The quantum well with the bigger atomic spacing than GaAs was the break-through that made mobile phones possible.

John Orton's definitive history reports that most of the development costs of the HEMT were paid for by the military. Once the transistor had achieved the performance the military required, it was public demand that drove commercial manufacturers to reduce the cost of mobile phones. As readers over 25 will be aware, mobile phone use took off at an exponential rate in the second half of the 1990s. That was just ten years after the groundbreaking research which I observed at Philips. Mobile phone technology has spread worldwide and, as we all know, has had a revolutionary impact in more ways than one.

Third-generation solar cells

We are nearly up to date with our semiconductor and solar cell history. We have met first-generation silicon solar cells and second-generation thin film cells. We still need to meet the third-generation solar cells, which could make a major contribution to the solar revolution in the future.

Third-generation mobile phones that connect to the internet have clearly enhanced communication and entertainment on the move. Third-generation PVs based, like third-generation mobiles, on GaAs technology offer many exciting possibilities because the cells are three times as efficient in turning sunlight to electrical power as first- and second-generation solar panels. Later, we will meet some novel applications of third-generation PV technology that could exploit this much higher efficiency.

The key to higher solar cell efficiency is that three different solar cells are grown on top of each other as one cell. This can be done with the similar technology to that used to grow GaAs mobile phone crystals. The top cell is made from a high band-gap alloy of GaAs, the middle cell is made of GaAs and the bottom cell is made of the low band-gap semiconductor germanium.

The highest energy, blue, green and yellow photons in sunlight are absorbed in the high band-gap top cell. The medium energy orange and red photons pass through the top cell and are absorbed in the GaAs cell cell and the low energy infrared photons are absorbed by the low band-gap bottom cell. Most of the photons in sunlight are absorbed in the cell appropriate to their energy. This means the photons have just the right amount of energy to raise an electron from the valence band to conduction band. The golden photons, which you will recall are the most plentiful in sunlight, are absorbed in the top cell, which is 'just right' to produce above 40 per cent efficiency.

To a physicist, the triple junction solar cell is an even more appropriate use of the name 'Goldilocks Principle' than the examples in astronomy, economics or psychology. Cinderella and Goldilocks; the history of solar cells is something of a fairy story. Part III will reveal if there is likely to be a happy ending.

The problem with the triple-junction cell is that it is still expensive. That is not a concern in space where it is now the solar cell of choice. On the ground, they have to be deployed in systems of lenses and mirrors, which concentrate sunlight onto small cells. This approach is known as concentrator photovoltaics (CPV). Most of the area is covered by cheap, plastic lenses or glass mirrors. The cell itself is typically one five-hundredth of the area of the lens so the cost is considerably reduced.

The standard triple junction cell has the same atomic spacing in all three cells. It turns out this means that the both the top and middle cell band-gaps are higher than is optimum for maximum sunlight harvesting. My year at Philips showed me the many benefits quantum wells were having for lasers. When I returned to Imperial I decided to see what advantage quantum wells might have for solar cells. It eventually became clear that their main benefit would lie in being able to lower the all-important band-gaps of the top and middle subcells *while still keeping a perfect crystal*.

Keeping the crystal perfect turned out to be a difficult problem for the middle cell of the triple junction cell. The only alloy of GaAs we could use to make a quantum well in this cell has atoms further apart than the atoms in GaAs crystals. It is easy to make one quantum well with that material. Keeping a perfect crystal with the 50 or more wells needed to absorb sunlight is a much bigger challenge.

With help from colleagues at the University of Sheffield and the University of Parma, we solved the problem by separating the quantum wells with an alloy in which the atoms are *closer together* than in GaAs. This meant they balanced the wider separation of the atoms in the wells. We were able to maintain a perfect crystal with 50 or more wells. Using this technology, the company QuantaSol which I co-founded achieved 40 per cent efficient triple junction cells, a comparable result to the market leader.

It was only when I read John Orton's history while writing this book that I realised the quantum wells in the QuantaSol cell were made from the same alloy as in the single quantum well in the HEMT that made the mobile phone possible. As we will see later, I believe these quantum wells can play as important a part in some novel future PV applications.

So let's move on to see what other renewable technologies are ready to help PV progress the solar revolution and why the revolution is taking off at such different rates in different countries. Then we will see how this third-generation PV technology might rise to the challenge of generating a solar replacement for our diminishing oil resources.

II

The Here and Now of the Solar Revolution

SEVEN

Questions for the Solar Revolution

Day and night, repay your loan, shine with sun's compulsive light.
Claire Crowther, *Mollicle*

In Part I we looked at how quantum ideas led to the semiconductor revolution and eventually yielded the first, fresh shoots of a solar revolution. The semiconductor revolution has made an impact on all our lives, but many readers may not be aware that a solar revolution is also underway.

The solar revolution is in fact growing very quickly in certain countries. Later on in Part II we will address the question of why it is developing at such very different rates in different countries. We will also meet a number of new technologies that you may not have been aware are powered by the sun. In addition, Chapter 9 will contain suggestions of ways you can lower your carbon footprint.

You may have concerns about solar power. So I will first describe some examples of how well solar technologies are faring in countries that are leading the revolution. I hope to answer the questions you have and those of sceptics who doubt the utility of solar power.

What happens when the sun doesn't shine?

The short answer is that the wind is usually blowing somewhere not far away. Yes, wind power *is* a solar technology. It is making a large contribution to the solar revolution in some countries. I want to show you how wind power works, so you can see that it is complementary to PV power.

Wind power results from the movement of air molecules in one direction. Remember that air consists of molecules of oxygen and nitrogen flying around at high speeds, randomly in all directions. Because molecules collide with each other very often, they don't in general move very far. I explained in Chapter 2 how the sound of a violin reaches your eardrum. The vibrations of a violin string pass through the air to your ear because air molecules rebound off other air molecules creating a wave. The air as whole does not move from the violin to your ear; the wave does.

Remember the dancers at the rave in Chapter 2? They transmit the gyrations of the drunken interloper across the floor, as a wave, by bumping their neighbours, while doing their best to stay dancing on the same spot. Wind is like the result of bouncers coming for the reveller. Neighbouring dancers would pretty smartly direct their steps away from the disturbance, forcing their neighbours to move away too. The dancers now move en masse to the quieter parts of the dance floor, keeping on dancing as long as the band keeps playing.

This mass movement of dancers across the floor is like the movement of air molecules forming a wind. The wind motion, and the back-and-forth motion of the molecules in a sound wave, are both superimposed on the random motion of the air molecules. Do you remember I said that there is no music on the airless moon? There is no wind either. Apollo astronauts had to remember to take a plastic Stars and Stripes flag with them.

What makes all the molecules in the air move in one direction?

Different air temperatures in different places are the primary reason. Imagine you are on a beach in the Caribbean. The air feels a lot warmer than in London, say, because the air molecules are moving faster. The higher the speed of the air molecules, the higher their energy of motion and the hotter the air feels, so the higher the temperature. It is similar to dancers at a rave feeling hotter the more energetically they dance around. We will not have to worry exactly how much faster air molecules move at higher temperatures. The exact mathematical relationship was first worked out by our old friend Maxwell. He inadvertently contributed to our understanding of wind power, as well as to PV power.

What is important is that sunlight does not heat the air directly. Only a small fraction of sunlight is absorbed by the atmosphere. Hence, on a clear day, a large proportion of the photons in sunlight reach the earth's surface. This is good news for PV systems and sunbathers on Caribbean islands. It also means that most of the photons in sunlight give up their energy when they are absorbed on the beach, in the sea (or on your skin, so be careful!). The atoms absorbing the $E = hf$ energy of the photons vibrate back and forth faster, so the temperature at the surface of the beach and sea rises. Some of this heat energy ends up being stored about two metres below the surface of the earth, which we will find useful in Chapter 9.

The jiggling atoms in the surface of the sand and the sea warm up the air by bashing air molecules so they fly around faster. This picture of the ground molecules warming the air molecules was like the one Lord Rayleigh used to try to predict the energy of electromagnetic waves in a cavity. Remember this idea was extremely misleading for light waves until Planck and Einstein sorted things out.

As the hot sand and sea warm up the air on your Caribbean island, the hot air expands and rises; just think hot air balloons. Air is drawn in from colder places, such as Europe. This is like air in a room forming a draft towards an open fire and up the chimney. Hence winds flow from Europe to the Caribbean. The hot air at high altitude flows in the opposite direction to balance the cold air flow. Hence winds tend to circulate between hot and cold regions.

Now other influences, such as the rotation of the earth, come into play, complicating the picture. Christopher Columbus and his crew were the first Europeans to explore the Caribbean thanks to the cold wind that flows towards the Caribbean to replace the rising hot air. They got back home to report their discoveries because their return voyages were always to the north of the outgoing one – a clockwise rotation. It is the same effect you can see on a TV weather map. Winds should flow from high to low pressure regions, like air flowing out of a punctured tyre. But because of the earth's rotation winds flow clockwise round the high pressure areas in the northern hemisphere.

The energy of motion of the air molecules in the wind can be turned into the energy of rotation of wind turbine blades. The turbine blades rotate coils of wire in a magnetic field. The rotating coils give us electricity according to Faraday's law. Not quite as simple as PV, but a lot simpler than controlling a nuclear reaction to produce steam to turn a turbine.

We can use this picture – that wind power is related to temperature differences caused by the sun at the earth's surface – to appreciate the complementary nature of our PV and wind power resources. Because sunlight heats the air molecules indirectly, it takes time for the air to heat up. The number of photons of sunlight hitting the earth is highest at noon when the sun is highest in the sky. Because the ground and buildings have to be heated before the air, the temperature of the air is highest in the afternoon. This explains why, in countries like the UK, wind power is, on average, highest mid-afternoon. This also explains why in Italy, California and many points south, electricity demand in summer is in general highest in the afternoon because of the air conditioning.

PV power supply peaks around noon every day, when the sun is high in the sky. On average, in the UK, wind power peaks in the afternoon. This is the first of a number of examples that show that wind and PV resources are complementary.

The peak electricity demand in the UK is at 7 p.m. in January. What use is solar power then?

As we have just seen: *it is precisely because it is dark and very cold in the UK at 7 p.m. in January, while at the same time the sun is shining brightly in places far to the south-west, that on average the winds in the UK are stronger on January evenings than in the summer.* As a certain British newspaper might claim, it's the Sun wot done it!

The sun makes the air in places like the Caribbean a lot hotter than in Europe on January evenings, but the difference is not so marked in summer. As a result, in the UK, and parts of Europe with similar climates, the wind power resource is higher in the winter than the summer. In contrast, the PV power resource is higher in the summer than the winter. In a paper in the journal *Energy Policy*, Massimo Mazzer, Kaspar Knorr and I looked at the way the electrical power demand in Germany and the UK varied during the year. By comparing this with the wind and PV resources averaged over each month, we argued that the UK will need less wind and PV power than Germany to achieve an all-renewable electricity supply. This is because wind and PV resources are complementary.

There are other ways that PV power is complementary to wind power. The further north one goes in Europe, the lower the sunlight resource averaged over a year. The closer you get to the oceans, the more humid and cloudy it gets and the lower the sunlight again. Wind is complementary. The further north in Europe and the closer to the oceans you get, the higher the wind resource on average.

In northern latitudes, the vertical east-, south- and west-facing walls of tall buildings are good locations for PV. Again wind is complementary. The roofs of these tall buildings are good locations for vertical-axis wind turbines; those elegant rotating spirals that are starting to be seen in our cities.

Is global warming due to human activity?

I hope you now appreciate how wind power and PV power are generated by our sun and also that they are complementary. Soon we will see how Germany is making good use of sunlight's power, which is loaned to all of us, day and night.

But first let's consider how to respond to those you may meet who say that a solar revolution is unnecessary. They maintain global warming has little to do with the carbon dioxide that human activity has emitted into the atmosphere. If you need a quick response, point out that over the past half-century the rise in global warming has broadly followed the rise in concentration of carbon dioxide in the atmosphere as we will see in Part III. You could then follow up with more detail on how the amount of carbon dioxide in the atmosphere and the earth's temperature are linked. We have met enough physics already for you to understand how this linkage works. It is crucial for the future of our civilisation.

This section is being written as the wind is lashing rain onto my window. The UK is experiencing what looks like being one of the worst winters for coastal storm surges and inland flooding in living memory. Now that you know how our winds are produced you should be able to appreciate that, because the average temperature of the earth surface has been rising for the past few decades, the number and severity of extreme storms, hurricanes and typhoons will inevitably increase.

The temperature of the earth's surface is determined by an extremely delicate balance between the number of photons of sunlight absorbed by the earth and the number of photons radiated from the earth's surface back into space.

QUESTIONS FOR THE SOLAR REVOLUTION

It is not widely appreciated, but the earth radiates photons just like the sun does. Remember the Victorian physicists who struggled to explain sunlight? The temperature of any solid is determined by how fast its atoms are jiggling around their fixed positions. The faster they jiggle the higher the temperature. Recall that Max Planck explained that the higher the temperature of any object the more the photons it emits and the more golden the colour of those photons. He achieved this without knowing *how* atoms emit photons.

According to Planck, you and I are emitting the number of photons appropriate to our body temperature. We are both very much cooler than the sun so the photons we emit are much longer wavelength than sunlight. They are infra-red, so you cannot see them. But you and I emit enough of them to set off a security alarm at night. If you are trying to explain this in a discussion with someone who denies global warming, they are probably emitting more infra-red photons than you as their temperature is rising with frustration. The expression 'incandescent with rage' is good physics.

For millennia, or at least back to the last ice age, the energy in the photons of sunlight hitting the earth has on average been equal to the energy of the infra-red photons radiated from the earth back out into space. Thanks to this delicate balance, the Earth's average temperature has remained remarkably constant for millennia.

Now something has upset this delicate balance. The earth's temperature is rising so either more photon energy is arriving in sunlight or less photon energy is escaping out into space from the earth. Fortunately, the temperature of the sun has remained constant for billions of years so the number of golden photons hitting the earth's surface has remained constant. So it must be that less infra-red photons are being radiated back out into space. What is happening to the photons?

The earth's atmosphere is like glass, it does not absorb many golden photons. But both our atmosphere and glass absorb a large number of infra-red photons. In the case of the atmosphere, this is because molecules of gases like carbon dioxide and methane are very good at absorbing infra-red photons. Even a very small amount of such gases can absorb many infra-red photons and stop them being radiated out to space. The infra-red photons absorbed by the carbon dioxide and methane molecules heat these molecules so they in turn radiate photons in all directions. Around half of the re-radiated photons return back to the earth's surface.

Glass is similar to air in letting sunlight through, but like carbon dioxide and methane it absorbs infra-red photons. About half of these will be radiated back into the greenhouse. This is why in summer it feels warmer in a greenhouse than outside. Hence the heating of the earth due to the absorption of infra-red photons is called the *greenhouse effect*. The highly absorbing gases like carbon dioxide and methane are called *greenhouse gases*.

As the amount of greenhouse gases in the atmosphere rises, the earth heats up. This is not only because fewer infra-red photons are radiated out to space but also because some of their energy is re-radiated back to earth to jiggle the atoms on the ground and in the sea again.

We have been disturbing this delicate balance since we started serious fossil fuel burning. As we will see in Part III, this balance is being even more seriously disturbed as the ice caps shrink. The exposed dark earth absorbs more sunlight than the shiny ice. There is less shiny snow to reflect sunlight directly back into space.

Even a small temperature rise means the atoms in the ground and the ocean are jiggling around more, so the air above is getting hotter. The air becomes less dense and rises faster. This creates faster winds like a hotter fire in a chimney. The winds in turn produce bigger storm surges on our coastlines. These excite the odd surfer but for most of us they strike fear for the future if the temperature of the earth continues to rise.

QUESTIONS FOR THE SOLAR REVOLUTION

Wind and PV are too intermittent to make a big contribution to our electricity supply without large amounts of battery storage

Wind power and PV power are intermittent sources of electric power – the power they produce varies with time – but they are intermittent in very different ways. Those who are sceptical of the renewables often claim that the intermittency of PV and wind power is a problem that will require large numbers of expensive batteries.

Let's look at a number of small experiments being made worldwide, and then one larger-scale experiment in Germany, to answer this question.

First, the small-scale experiments. Worldwide, an increasing number of householders, are conducting experiments to see if a PV system on their roof can supply a significant proportion of their electricity requirements for 25 years without *any* battery. If you, dear reader, are such a pioneer – well done and welcome to the revolution! You will probably have a connection to the national grid through one or two electricity meters. When the sun shines the solar electric power from your panels can be used to run household electrical equipment. Any excess electrical power is fed into the grid. When the sun goes behind the clouds, and at night, this same equipment runs on power from the grid. So the grid is acting rather like a battery, taking your excess electrical power and putting it to use elsewhere, then providing you with power when there is no sun.

Most of these experiments are successful and householders are seeing electricity bills that are somewhere between 25 per cent and 50 per cent lower than previously, depending how often they need electricity when the sun shines.

In addition, as I discovered when we recently installed PV at home, it is nowadays easy and cheap to store excess solar electricity for use when the sun is not shining. In Chapter 9 we will find out more about a small electronic 'black-box', which allows any excess electric power to trickle into the immersion heater in a hot water tank. This means the excess solar electrical energy is being stored as heat energy.

Our PV was installed in summer. When, at the end of September, we switched on the gas central heating we found we did not need to switch on the gas hot water system. In August and September, we only switched on electric power from the mains to the immersion heater when we had visitors to stay (we are a two-person household). In October, we turned on the immersion heater from the mains on seven particularly cloudy days. We are therefore expecting not to have to use gas to heat our water for about two-thirds of the year.

The large-scale experiment, which is little known but extremely important, started in Germany on 1 January 2006 and ran for the whole year. The project was called Kombikraftwerk, which is German for 'combined power plant'. A computer program matched 1/10,000 of the actual, real-time electrical power demand on the German grid using the actual real-time power output of a number of wind, PV and biogas electricity generators. The experiment worked. The electrical power demand matched the electrical power supply throughout the year.

Those who are sceptical of the contribution the renewables can make often claim that if we try to use only renewable energy 'the lights will go out'. That was not the case at any time that year on the Kombikraftwerk experiment.

You can see the actual German demand and the way it was met by the solar supply for a typical week in September 2006 in our *Energy Policy* paper. I have studied the German demand figures on the German grid for every day in 2006 and compared them with the demand figures in the UK in 2010. They are quite similar. The way the demand varies through a working day, in both countries, is what you might expect with a moment's thought. There is a steady rise in the morning as homes, workplaces and public buildings start up. The peak demand is around noon, followed by a much slower fall in demand in the afternoon. There is usually a second peak in the early evening as family meals are prepared. This evening peak gets higher in winter until, around the turn of the year, it is above the daytime peak in both countries. Germans cook food and watch television in much the same way as British citizens. The Kombikraftwerk experiment showed the actual output of wind and biogas generators was able to cope with these 7 p.m. winter evening peaks.

The way the wind and PV electrical power supplies in Germany vary over each day, during the typical week and also over the year is very interesting. Yes, wind and PV supply are both variable, but in very different ways. The wind contribution is intermittent, changing up and down in a short time. In addition there are periods of a few hours where the wind is very high or very low. The Kombikraftwerk team found these periods could be predicted quite well from weather forecasts.

You can see from the typical week shown in the *Energy Policy* paper that the biogas electrical power contribution goes up and down rather like the wind. This is because the computer running the Kombikraftwerk experiment was adjusting the biogas power contribution to compensate for the changes in wind power supply. The biogas electricity generators used in this demonstration could increase and decrease their output in minutes.

The variability of the PV supply contribution was completely different. Every day of the year, including winter, it rises to a peak around noon, when the sun is highest in the sky, then falls back in the afternoon. The height of this peak varies day to day depending on the amount of cloud cover, which, like the amount of wind, can also be forecast quite accurately. The PV power contribution from the actual PV generators in Germany therefore varies with time in a way that is predictable.

Most importantly, in summer and in winter, the solar power

contribution peaks within an hour or so of noon. That is around the time when the electricity is most needed: the time of peak daytime electricity demand in both Germany and the UK.

Another very significant result from the 2006 experiment is that, over the year, wind and PV generators supplied a very impressive 78 per cent of the power. The main backup was provided by the biogas generators, but they only had to supply 17 per cent of the power. The Kombikraftwerk experiment did have some hydro-power storage. Electric power was used to pump water back up behind a dam to store the excess electrical energy as the energy of position of water. When backup electrical power was needed this energy stored in the water was converted into the rotational energy of the turbines and thence to fresh electric power. Over 2006 only 5 per cent of this storage power was required by Kombikraftwerk. In fact, more excess electrical energy was stored behind the dam in the year than was released to backup the other renewables.

The Kombikraftwerk experiment clearly disproves the claim of the solar sceptics that massive amounts of storage will be necessary to backup large contributions of renewable power. The scale-down demand of the German grid was supplied in real-time throughout the year by a power supply that was 95 per cent renewable: only 5 per cent storage power was needed.

That is not to say that solar electricity would not benefit from more storage. What the Kombikraftwerk experiment shows is that we can move fast on installing new PV and wind power systems without waiting for improvements in storage options as long as we have biogas generators.

Solar electricity is too expensive

Another commonly held view is that PV electricity is too expensive. Some analysts claim it will be many years before the price of PV electricity in the UK achieves *grid parity*. This is when the price of electrical energy supplied to the grid by PV generators is comparable with the price from conventional sources.

In fact, PV electricity has already achieved grid parity in Italy and Germany. Wednesday 2 May 2012 was an important day in the history of the solar revolution. It might well have been overlooked had my old friend Massimo Mazzer not been studying data on electricity generation in Italy for our paper. At 2 p.m. on that day, the price of electrical energy on the southern Italian grid did not just drop below the price of conventional electricity; it *fell to zero*. This is particularly fortunate for Italy. As I explained earlier, electricity demand is high on sunny afternoons in countries like Italy due to the use of air conditioning.

Thanks to PV, on sunny afternoons in southern Italy electrical energy is now free. It is 'too cheap to meter'. This was the prediction infamously made about nuclear power in its early years. Nuclear has never achieved it, and, I confidently predict, never will. Ironically, it has become reality for PV in Italy at a time when the cost of nuclear electricity is going off scale in the opposite direction. This is because of the safety implications of the Fukushima disaster and the delays in building prototype reactors in Finland and France.

Some might assume this could only happen in sunny southern Italy, but let's look for evidence from Germany, which has significantly less sunshine and significantly more PV.

At noon on 2 June 2011 two more important events happened, this time in Germany. First, the PV power supplied to the German grid reached 28 per cent of the demand. Second, for the first time ever, the wholesale price of electrical energy supplied to the German national grid fell below the night-time price. The two events were not coincidences. The PV power caused the fall in the price, as I will explain.

The wholesale price of electrical energy on any national grid is higher during the day, when demand is high, than at night when demand is low. In Germany in 2007 the peak daytime price was

30 per cent higher than the lowest price at night. By spring 2011 this had dropped to 10 per cent higher.

We know that cheap PV power produced this significant price fall for three reasons. First, the peak price of electricity on any national grid occurs at the time of maximum electricity demand. In Germany, as in London, the maximum daytime demand occurs near noon as noted in the last section. Also, the peak PV supply occurs near noon when the sun is highest in the sky.

Second, the amount of PV power connected to the German grid rose from 4.2 GW to over 17.4 GW during the four-year period that the price difference was dropping from 30 per cent to 10 per cent.

Third, as you can see from a graph in our *Energy Policy* paper, this price difference was higher around 1 January than around 1 June each year. This is to be expected if PV is responsible for the price fall. The sun is lower in the sky at noon in winter than in summer. But the wholesale price difference, day to night, around 1 January 2008 was above 30 per cent, whereas by 1 January 2011 this price difference had fallen to 15 per cent. Even in winter, the wholesale price of electrical energy has been falling as the amount of PV power on the grid has been increasing.

The important conclusion from the fall in the wholesale price of daytime electricity in Germany is that PV electrical energy in Germany has not just achieved parity with all other supplies on the grid; the price is *lower* than any other type of electricity on the grid. The PV power supplied to the German grid around noon is cheaper than any other type of power *at that time*. Noon is not just anytime. It is the time of maximum daytime electrical power demand.

So how come some analysts maintain that grid parity is many years away in the UK? After all, averaged over the whole country and the year, the photons in sunlight contain only about 5 per cent less energy in the UK than in Germany.

Such inconsistencies between projections and reality arise because of the confusion between energy and power, which we met earlier. Remember that *energy* is what your favourite footballer has given to the ball which hits the back of the net. *Power* is all about the timing and the force with which he or she kicked the ball. In the grid-parity discussion, the confusion arises because it has been traditional for electricity companies to charge us for the energy supplied rather than power supplied. The good news about the fall in the peak wholesale price of electrical energy in Germany is that what matters is grid parity in terms of the price *at the time that the electrical power is supplied*. The power and the timing of the kick are crucial if you want to give a football high energy of motion. The price of PV energy supplied to the German grid is clearly lower than the price of conventional electric energy at that time, because the peak price is falling. PV power arrives near the time of peak demand when the cost of conventional forms of electrical energy is highest. With only 5 per cent less solar photons each year, I suspect that grid parity at the time the PV power is supplied has already been achieved in the UK.

The importance of power rather than energy in determining the price of electricity on the grid was clearly demonstrated by the important event in Germany on 2 June 2011. Remember that PV supplied 28 per cent of the *power* on the German grid. As a result the daytime peak price of electrical energy fell below the lowest night-time price for the first time. Averaged over the year 2011, however, PV only supplied 3 per cent of the electrical *energy* on the grid. If it was energy that mattered rather than power, a mere 3 per cent would not have made such an impact.

As far as I am aware, there were no major problems when PV supplied 28 per cent of Germany's electric power requirements. That observation is relevant to another question you may well be asking.

Can national grids cope with large amounts of wind and solar power?

Many people are concerned that national grids will face operational difficulties when wind and PV supply large amounts of electrical power to the grid. Over the years, there have been a number of studies suggesting the UK grid could not cope with wind or PV supplying 20 per cent or so of the power. The experience from Germany of large amounts of wind and PV on the grid is very reassuring on this question.

When, around noon on 2 June 2011, PV supplied 28 per cent of the power demand on the German grid, the only consequences were positive: the peak price fell below the night-time price. At noon on Saturday 26 May 2012 PV contributed close to 50 per cent of the power demand on the German grid, again without problems.

The German grid has also coped with large amounts of wind power. For 127 separate one-hour periods between 1 June 2010 and 9 March 2011, wind power successfully contributed over 30 per cent of the supply to the German grid.

These achievements in Germany are reassuring. But it is clear that improvements to all national grids would help solar technologies. Recall that the price of electricity in southern Italy now falls to zero in the early afternoon on sunny days. This is not just due to PV; it also is partly due to bottlenecks in the national grid. These make it difficult to export excess electricity up the boot of Italy. Improvements to the Italian grid would mean that excess electrical power from the south could be exported north and reduce electricity prices elsewhere. Also, the excess electrical power from the south could be stored as the energy of position of water behind the many hydropower dams in the Alps in northern Italy.

Compare this with Britain's plans for new nuclear build. If the French government owned company EDF does build a 1.6 GW reactor at Hinkley Point in Somerset, the UK National Grid will have to spend over £1 billion on improvements to the grid to make use of its power. This means bigger electricity bills for five years during the building of the reactor followed by further price hikes for the consumer due to the subsidy which will be payable should the reactor eventually work.

By contrast, as PV on buildings and community-owned wind and PV farms develop, so could small local grid networks. They would be smaller, cheaper and much less obtrusive environmentally. The growing number of community owned renewable power schemes, which we will meet in Chapter 9, are likely to find it convenient to develop their own local networks. It makes sense to share electrical power locally. Just think of the differences in electricity demand patterns between homes on one hand and schools and offices on the other. If these local grids also had some biogas electricity generation the local grid could be nearly independent of the national grid as in the Kombikraftwerk experiment. These local grids need only have a few points of connection with the national grid. This would also insulate the local grid from power cuts and likely price rises on the national grid.

In October 2012, Kaspar Knorr (the current leader of the Kombikraftwerk project), Massimo Mazzer and I summarised data on the performance of PV and wind on the German and Italian grids, and the results of the Kombikraftwerk experiment, in the scientific journal *Nature Materials*. We concluded that PV, wind and biogas electric generators are so effective that a government could impose a moratorium on building any new electricity generators other than renewable ones. This would be the most effective way to lower carbon emissions worldwide. The evidence from Germany and Italy suggests the electric power would be cheaper and there should be no need to fear that the lights will go out.

How fast can PV and wind power expand?

In some countries PV installations have been expanding very fast indeed. When Kaspar, Massimo and I made a graph of the cumulative power of PV installed in recent years in Germany, Italy and the UK we had to use a special type of scale (which rises 1, 10, 100 … rather than 1, 2, 3 … as on a normal graph) to show all the data points properly. You can see it in our *Energy Policy* paper. Between 1996 and 2011 PV installations in Germany increased on a fairly straight line on our graph. That means the expansion in Germany in those years was *exponential*, which is a very fast increase indeed.

Some of you will remember how fast the mobile phone revolution took off in the late 1990s. It too was exponential. In 1995 only the police, top executives and celebrities had mobile phones. By the turn of the century nearly everyone seemed to have one. In fact, the average increase in PV installations in Germany in the five years to 2010 was higher than the average increase in the number of mobile phones in Western countries over the five years to 2000.

It was a remarkable achievement by Germany to maintain an exponential increase in the amount of PV power installed for a decade and a half. This was mainly thanks to their feed-in-tariff (FIT) policy. This is a guaranteed price for the electrical energy from PV, which is fed back into the grid. The guaranteed price was set to fall in a prearranged fashion known to all. In that way, the new PV companies could plan their investment strategies. The tariff is paid for by a levy on the electrical energy purchased by most electricity users in Germany. The aim was to create a market leading to mass production and falling prices. In this, the scheme was successful. As we will see, Chinese industry was, however, the main beneficiary as they invested in larger-scale manufacturing and stronger supply chains.

The German FIT has been so successful that in recent years the price of PV panels has been falling faster than expected. From time to time the German government cut the price paid to PV generators by more than the original planned gradual reduction. They have managed these extra cuts in a way that has maintained a steady exponential increase. I am apprehensive about the effects of the recent round of extra cuts in Germany, particularly as European PV manufacturers are under pressure from cheap Chinese imports.

QUESTIONS FOR THE SOLAR REVOLUTION

The graph that Kaspar, Massimo and I published shows the dramatic effect of the stimulation policies belatedly introduced in Italy and the UK. From 2006 to 2011, Italy expanded exponentially at a faster rate than Germany so that they came close to catching them up, but their tariff was then drastically cut and the exponential rise stalled. The UK introduced a FIT similar to the German one in April 2010. At that time, Germany had nearly 250 times as much PV as the UK. In 2010 and 2011, the new PV installations in the UK, as in Italy, expanded faster than in Germany.

The British version of the FIT differed from the German one in a number of important respects including there being no gradual tariff reduction planned. In late 2011 the government announced a draconian cut of 50 per cent, far larger than anything Germany had implemented and a severe blow to the fledgling PV industry in the UK. Despite this, instead of having 250 times as much PV, the German lead had fallen to 17 times as much as Britain by the end of 2012.

How fast can PV expand in the future? How soon could Germany, Italy and the UK have the PV contribution required for an all-solar electricity supply? In the Energy Policy paper we showed how much PV each country would install if they introduced a FIT like the German one, at a level appropriate to current PV prices, so that installations expand at the average rate Germany actually achieved in the five years from 2006 to 2010.

The Kombikraftwerk project showed that the PV contribution to an all-renewable electricity supply in Germany should be around 55 GW of PV power. If German installations continue to expand as they did from 2006 to 2010, they should reach this target by 2015. The Italian national electricity demand is lower than in Germany, but Italy has more sun. If Italy reintroduced a FIT that gave them an expansion rate closer to Germany, they too would achieve the 55 GW German target well before 2020.

We saw earlier that PV and wind power in the UK are complementary. As a result, the all-solar target set by the Kombikraftwerk

163

project for Britain is only 37 GW. This is a reduced PV target compared to Germany or Italy because the average monthly variation of PV and wind power is more closely matched to the UK electricity demand. Even though the UK started well behind the other two countries, were a FIT set at an appropriate level like the German one, Britain would have sufficient PV for an all-renewable electricity supply by 2020. The projected 37 GW of PV would be far higher than the UK government's optimistic expectation of 1.6 GW of new nuclear build by 2023.

How fast can wind power expand? Wind power took off exponentially in Germany from 1995 to 2002 when the rate of expansion slowed. One reason why, unlike PV, wind power did not continue to expand exponentially is that wind generators are much larger than solar panels. The smaller the size of an electricity generator, the easier it is to mass-produce and the quicker it is to install. Instead, wind turbine and wind farm sizes have increased over the years because that makes them cheaper for each electrical energy unit they produce. And the bigger the wind turbine, the stronger the objections during planning applications.

The UK was around nine years behind Germany in onshore wind installations in 2006. In that year Massimo Mazzer and I calculated that if the UK installed onshore wind at the rate Germany had achieved in the previous decade, Britain could have 25 GW of onshore wind by 2020. Sadly the UK has fallen even further behind. By 2012 Britain had only achieved 6 GW of onshore wind. Germany took nine years to move from 6 GW to 25 GW. So if Britain were belatedly to implement German policies in 2014 they could have 25 GW of onshore wind by 2023.

The UK is fortunate in having a much longer coastline than Germany and hence a much larger offshore wind power resource. The UK government has belatedly recognised this and is aiming for 33 GW of offshore wind by 2020. Allowing for three years delay, the total wind power generation in the UK in 2023 could be as high as 58 GW. Like my PV prediction, this would dwarf the British government's very optimistic assumption of 1.6 GW of new nuclear build by that year.

A wind power prediction of 58 GW by 2023 is over halfway towards a target for wind power in the UK of 104 GW. This is the target, which the Kombikraftwerk experiment suggests, would give the UK an all-renewable electricity supply. Like the PV target above, 104 GW takes into account that wind power in the UK has an average monthly resource variation more closely matched to the British electricity demand than the German.

Hence simply by learning from the experience in other countries while sticking to its own plan for offshore wind, the UK could have *all* the PV power and more than half the wind power required for an all-solar electricity supply by 2023.

Many other renewable electricity-generating technologies are already available and contributing to the solar revolution. There are so many I will need the whole of the next chapter to describe them.

EIGHT

The Solar Cornucopia

There is no new thing under the sun.

Ecclesiastes 1:8

In the last chapter, we saw that wind and PV power are complementary electricity-generating technologies. They are also capable of the fastest growth. Together with biogas they can satisfy most of the electric power demand of Germany and many other countries including the UK. The evidence from Germany and Italy shows governments that only wind, PV and biogas are necessary to supply national electricity demand.

This chapter should further reassure any government thinking of going for an all-renewable electricity supply. There are many more solar technologies ready to be deployed, which will mean an all-renewable electricity target is achieved more quickly. Some of these technologies may well be necessary in countries with solar resources very different from Germany.

To misquote solar revolutionary Al Gore, now is the time to face the 'convenient truth' that we are blessed with a glorious plethora of solar technologies that can help achieve all-solar electricity generation. This chapter will describe some of them briefly before we go on to compare their carbon footprints. In the next chapter we will see how householders and local community groups can make use of these technologies to generate their own electricity and to reduce the amount of natural gas they burn.

I apologise if I have not given sufficient prominence to your favourite renewable technology. They will all have a part to play in the solar revolution.

Ecclesiastes was right in that many of these renewable technologies have been around for a long time. Where Ecclesiastes was wrong is that some solar technologies are made practical and low carbon thanks to the existence of semiconductor technology. You now know this was new under our sun in 1947.

Solar hot water

Also known as *solar thermal*, this technology, like PV, has small- and large-scale versions. Small-scale, rooftop solar hot water systems are an established technology, much in evidence in the Mediterranean region. It has great potential for reducing the use of natural gas and electricity for domestic hot water in most parts of the world. New versions also can be used on east-, through south- to west-facing vertical walls in northern latitudes. A solar thermal system is much cheaper than PV, but you may need to fit a new, large hot water tank to make use of it.

Solar thermal, like roof-top PV, is a small-scale technology and so with the right stimulation package has the potential for exponential expansion. Where a number of solar technologies are concerned, Schumacher's principle can be extended: 'Small is beautiful ... and small can expand fast'.

There are some alternative approaches for domestic water heating that are new on the market so don't type 'solar thermal price comparison' into your search engine until you have read this and the next chapter.

Concentrating solar power (CSP)

CSP is the large-scale version of solar thermal. In this case, Ecclesiastes was spot on. Legend has it that more than two millennia ago the Greek scientist Archimedes got soldiers to hold their polished shields so they focused sunlight onto invading Roman ships. Apparently they succeeded in setting some of them alight. CSP is an extreme example of solar technology taking a long time to spin off from a military application.

Nowadays mirrors concentrate sunlight onto tubes filled with water, or a heat storage medium. If water is used and heated enough, the water boils and the steam drives turbines, which generate electricity. Like concentrator photovoltaics (CPV), the first application of CSP is large-scale electricity generation in desert regions. I will discuss this application further in Part III. CPV and CSP have complementary advantages in deserts. CPV wins in terms of electrical efficiency and ease of modular construction. CSP wins if storage of heat and electrical energy are required.

Large-scale hydropower

I explained in Chapter 2 how hydropower won the race to generate the world's first commercial electricity in 1881. I also explained the basic physics in Chapter 6. Hydropower already makes a major contribution to electricity generation in mountainous countries such as Italy, Norway and Switzerland. Solar sceptics claim that the best locations for hydropower are already taken and new ones will be unpopular with environmentalists. Remember, however, that existing dams can contribute to the solar revolution by storing excess renewable electric power as energy of position of the water behind the dam.

THE SOLAR CORNUCOPIA

In the last chapter, we saw that the price of electricity has fallen to zero on summer days in the early afternoon in southern Italy. Upgrading the Italian grid would mean the excess PV power in the south could be stored behind the many dams in the northern Alps and then run through the turbines when demand required.

Japan has built an excess of hydropower storage. That was done some years back when it was expected that nuclear power would eventually dominate the country's electricity supply. A dominant nuclear contribution would mean Japan had excess electric power overnight when demand is low. Once working, the power output of most nuclear reactors has to remain constant. Then came the Fukushima accident. Ironically, this hydropower storage will now be very useful. It can store excess wind and PV electrical power if, as seems likely, Japan replaces its unpopular nuclear plants with renewables.

Small-scale hydropower

There is still a large untapped potential for small-scale hydropower applications in rivers and fast-flowing streams. They need not be environmentally intrusive. There is one on the River Frome in Somerset near to where I live. There is absolutely no indication that electricity generation is taking place in the quaint, country building until one walks close enough to read notices.

Small-scale hydropower, solar thermal power and small vertical-axis wind turbines on tall buildings could all expand fast, as PV has, with appropriate government support. I gather from a friend that considerable bureaucracy is involved in getting a small hydropower turbine connected to the grid. Again there is much to learn from Germany where more of the onus for grid renewable grid connections is on the electricity supplier.

Geothermal power

This is yet another renewable technology that can be large scale and small scale like PV, wind, solar thermal and hydropower. We will meet the small-scale version in the next chapter.

We have met some of the physics already. In the last chapter, I explained that most of the photons in sunlight end up heating the earth's surface, rather than the air, with their $E = hf$ energy. That means the atoms just inside the earth's surface are jiggling about their fixed positions faster, so their temperature is higher than the air. As a result of this bombardment of sunlight, there is a region of the earth about two metres down where the ground is relatively warm. This is very important for the ground source heat pumps we will meet in the next chapter.

As you go further down into the earth, the temperature keeps on rising. This is mainly due to energy from Einstein's equation $E = mc^2$. The decay of radioactive nuclei in the earth's crust liberates radiation in the form of alpha particles, electrons or high energy photons (see Chapter 4). These energetic particles bash into the nuclei of atoms in the rock. The atoms therefore move back and forth faster about their fixed positions in the rock. Hence underground rocks are at a higher temperature than at the surface.

Even further into the earth the temperature continues to rise approaching the earth's molten core. The red-hot core is the remnant of the fiery birth of our planet from the collapse of interstellar dust. Fortunately, the primeval earth was not big enough for the collapse to lead to fusion reactions as in the sun. However, the molten core is still very hot, four billion years later, because the rock and soil in the earth's crust does not conduct heat energy very well. That is why solar energy is stored so well two metres in from the top surface.

The heat from the molten core and from the radioactivity in the rocks flows slowly up to the surface and helps maintain the relatively high temperature not far below the earth's surface. In some places, where the crust is thin or in volcanic regions, a higher heat flow reaches the surface. Examples are the geysers of Iceland (where geothermal energy makes a major contribution to electricity supply and heating buildings), and, much nearer to my home in Somerset, the hot spa water in Bath's Pump Room (where Claire and I got married).

Modern technology makes it possible to drill a well several kilometres down to regions where the temperature is far above 100 °C. If the geological surveys were correct, water and steam will gush up out of the well. They can in principle be used to drive a turbine and generate electricity. Usually a second hole is drilled and the water is pumped back to replenish the reservoir and maintain control of the extraction.

That sounds fairly straightforward once the holes have been dug. The system provides a constant, predictable, source of renewable, low carbon electrical power, which does not vary with time like wind and PV. What is not to like? Here lies another mystery. Geothermal energy shares a distinction with PV. Neither appeared in the list of top eight renewable technologies on the road map published by the British Department of Energy and Climate Change (DECC) in July 2011.

I've been doing some detective work, searching the pages of *Sustainable Energy – Without the Hot Air* by Professor David MacKay, who is the chief scientific adviser to DECC. I think I have found a clue to the solution of this mystery. On one page MacKay calculates a 'sustainable-forever' upper limit on the geothermal energy extraction from the optimal depth that is *lower* than the number he quotes on the previous page for the geothermal heat flow at the surface of the earth.

That sounds suspicious. Let's look closely at MacKay's assumptions when he calculates a 'sustainable-forever' upper limit. You can join in with an experimental test suggested by Professor MacKay in his book.

MacKay proposes that extracting geothermal energy is like 'trying to drink a crushed-ice drink though a straw'. If your initial rate of sucking is not sustainable you find you are soon sucking air. As he says, 'You've extracted all the liquid from the ice around the tip of the straw'.

You could test MacKay's idea next time you have a crushed-ice drink. I think you will find it is possible to drink it more sustainably. I suggest you hold the straw about a centimetre into the crushed ice and suck hard. When you find you are sucking air, ignore the embarrassing glances from your friends at the sound you emitted and hold the straw in place for a few moments. Very soon you will see the surrounding crushed-ice drink has flowed back around the tip of the straw. You can then suck more gently, keeping the tip of the straw moving down slowly to finish your drink more sustainably.

This flow back of the liquid is important. We all know that if you wait a few minutes after finishing, enough drink will have trickled down for you to have a few more gulps.

It seems to me that Professor MacKay has chosen not to include the drink that drains back in his calculation. If that is so it might explain why his 'sustainable-forever' limit seems to be so small. The analogy with heat energy underground is that as soon as heat energy has been extracted, so that the temperature at the tip of the borehole falls, the surrounding rocks will be at a higher temperature. Heat energy will therefore flow towards the tip. How fast heat energy flows back will fix how fast the heat energy can be extracted sustainably.

MacKay appears to be assuming that the 'sustainable-forever' rate of heat energy extraction is given by the geothermal heat energy welling up immediately under the borehole. His assumption is like having the iced-drink in your glass being constantly refilled. If you suck through the straw at the same rate as the drink is replenished the 'sustainable-forever' rate does indeed become the rate at which the drink is fed into the glass.

My view is that the 'sustainable-forever' limit should include the heat which drains back around the borehole when the temperature falls. The calculation should not just include the heat welling up immediately below the borehole but also the heat welling up some way away from the borehole. This heat will flow sideways to the tip of the borehole as soon as the temperature starts to fall. This corresponds to sucking from a large bowl in which the drink is being replenished at the same rate all over the top surface. Such a calculation should give a higher 'sustainable-forever' limit than one based on MacKay's assumptions.

The actual limit in a geothermal borehole will depend on how fast the heat energy drains back around the tip of the borehole through the nearby rocks. I mentioned that heat energy flows quite slowly through rock. In the Bibliography you find a paper by Ladislaus Rybach, which was published in the *Geo-Heat Center Bulletin*. This gives a full calculation of a 'sustainable-forever' limit. It supports my view that heat flows back around the borehole tip sufficiently fast that larger amounts of geothermal energy can be extracted sustainably than under MacKay's assumptions.

The Rybach paper first describes a situation where the extraction is so fast that the heat energy is depleted in 20 years. If the extraction then stops, the temperature at the tip of the borehole starts recovering quite quickly. The paper then looks at what happens if heat energy extraction is reduced to *one-quarter* of the first, unsustainable rate. As soon as the temperature around the tip of the borehole starts falling, heat energy starts flowing towards the tip. The quick recovery due to heat flowing back is happening *all the time* while heat is being extracted more slowly.

What Rybach's calculations show is, not only can power extraction continue at the sustainable level *after* 20 years, but the total heat energy extracted over the *first* 20 years is more in the sustainable calculation. The author concludes that sustainable power extraction provides more heat energy than the unsustainable approach *even in the first 20 years*. Such a result would not be possible on MacKay's assumptions.

I believe MacKay's sustainability calculation is too pessimistic. I was therefore pleased to learn that the Renewable Energy Association had commissioned a report by consultants Sinclair, Knight and Merz on the geothermal energy potential in the UK. You can find details of this report in the Bibliography. The report concludes that geothermal generation energy could provide 20% of the UKs electrical power supply. In addition the heat energy could supply the total annual heat consumption in the UK. This is our first example of what is known as combined heat and power (CHP). As well as electricity for the grid the geothermal stations could provide hot water for domestic and industrial central heating by a network of pipes, a system already much used for waste heat from electricity generators in Denmark.

The final word on the sustainability of geothermal power has to be experimental evidence. The first geothermal steam engine was demonstrated in 1904 in Larderello in Italy. The region is still producing geothermal energy a century later. The Roman baths in Bath are still being heated, two millennia after their construction. Yet again, Ecclesiastes was right.

I said that much of the exploitable geothermal energy is due to radioactivity and hence to $E = mc^2$. Should I be describing geothermal power as a nuclear technology rather than a solar one? It would be hypocritical to exclude geothermal energy from this chapter on the grounds that much of it originates from radioactivity. As nuclear supporters often point out to me in debates, my own beloved PV energy originates in the sun from $E = mc^2$.

It is also worth thinking a bit more about the nuclear fission reactor that nature has arranged for us deep inside the earth's crust. The earth's crust is thick enough that the heat energy flow from the primordial core is reduced so that the molten core is not cooling too fast. Also, the thick crust ensures that most of the nuclear radiation from deep in the earth doesn't reach the surface.

Hence the radiation at the earth's surface from the underground radioactivity is not so high that it threatens life on earth.

Nature has given us such a beautifully balanced fission reactor deep inside the earth that I am surprised that one of the originators of the Gaia hypothesis, James Lovelock, author of *The Revenge of Gaia*, should be so keen on new fission and fusion reactors. Nature's geothermal fission reactor can contribute much more to our electricity needs more quickly and more safely than new nuclear. Furthermore, the *fusion* reactor at the centre of the sun provides us with more than enough photons for all our energy needs. Deep in the earth and deep in the sun are the safest places for fission and fusion reactors respectively. We do not need any on the earth's surface.

Wave power

The tale of Salter's Duck is a salutary one for renewables in the UK. It is described by David Ross in a paper in the journal *Science and Public Policy,* which is referenced in the Bibliography. Ross maintains it was an early warning of the power of the nuclear establishment, and fossil fuel interests, to influence the direction of renewable energy developments in the UK. The Duck, which harnesses the energy of motion in sea waves, was invented by Professor Stephen Salter of Edinburgh University in 1972. A UK wave energy research programme was started in 1978 and ran for six years. At the time renewable energy research in the UK was supervised by a unit within the United Kingdom Atomic Energy Authority (UKAEA). In 1982 a decision was taken that the Duck was not economically viable and funding was withdrawn. Ross relates how the manager of the wave energy programme was excluded from the meeting which made the decision.

Thirty years later, the UK has at last woken up to the fact that it has a third of Europe's wave resource. The Renewable Energy

Association reports that two Edinburgh companies are leaders in wave power technology. Pelamis exported the world's first array of wave energy generators to Portugal. Aquamarine Power's Oyster wave technology has been delivering power for more than two winters in the Atlantic off the Orkney Islands.

Tidal power

It is well known that the moon and the sun both influence the tides. Hence tidal power is not strictly 100 per cent solar energy. However, it is renewable and can provide plenty of low carbon electrical power, particularly from islands. Like geothermal energy, it has the advantage of predictability, but it is not as constant a resource. Again the UK has belatedly woken up to its potential: the UK has half of Europe's tidal energy resource. A UK company, Marine Current Turbines, has installed the world's first tidal stream power station, SeaGen, in Strangford Lough, Northern Ireland.

I am a big fan of all renewable technologies, but I have a particularly soft spot for tidal power. I inadvertently contributed to one approach to harvesting tidal power. I was giving an undergraduate lecture course that included some hydrodynamics – the study of the way water flows. I tried very hard to make the problem sheet interesting as many physics students consider the subject old fashioned and boring. It was early November – the time in the UK for bonfires and fireworks. I set a problem about a box of fireworks being ignited accidentally and resourceful physics students trying to douse the explosions by squirting a jet of water from a water butt.

I was not prepared for the reaction the new problem sheet stimulated. First, I had a visit from a tutor who maintained I had made a mistake as my answer was inconsistent with the law of conservation of energy. A second tutor arrived to report that he and his students had tested the idea with water in paper cups with holes. They could see that, in certain conditions, the theory worked.

THE SOLAR CORNUCOPIA

A third tutor, Geoff Rochester, rushed into my room.

'I know … I know … I got it wrong … I apologise,' I stammered.

'What do you mean?' Geoff replied. 'Do you realise that if you had an underwater tidal stream and you inserted a tube which kept the tidal flow rising to the surface then, at the surface, the water in the tube would be moving with the speed the tidal stream had below the surface. The flow from the tube could turn a turbine and generate electricity!'

'Er, no … I hadn't realised that,' I confessed, 'Can you go over that again?' We sat down together and, yes, Geoff was right and so was my question – the conservation of energy was respected. The horizontal energy of motion of the water flowing at depth can be translated into the vertical energy of motion of the water in the tube at the surface of the water. This is because the extra pressure the tidal stream experiences in the deep is exactly cancelled by the change in energy of position as the water flows up the tube. As long as the water flows smoothly, conservation of energy ensures that the energy of motion the water had in the tidal stream is the same as it has on emerging from the tube at the surface of the water.

Geoff founded a company, HydroVenturi, with John Hassard, a colleague from my particle physics days. Yet again there is nothing new under the sun (or under the water). The Italian physicist Giovanni Venturi had beaten all three of us to the idea by 300 years.

The great advantage of this hydropower approach is that there are no moving parts underwater to go rusty. The problem turned out to be how to keep the water in the tube flowing smoothly. One clever idea Geoff and John came up with was to have a flow of gas passing with the tidal stream to take the energy of motion to the surface.

Unfortunately, there is nothing new under the sun where the fate of novel solar technologies in the UK is concerned. The HydroVenturi technology has yet to be exploited for electricity generation. John has reassured me that a major technical challenge has been overcome. All being well, yet another British marine energy technology will arise again from the depths.

Electricity generation by biogas and biomass

In the last chapter, we discovered that biogas electricity generators proved to be the ideal backup to wind and PV in the Kombikraftwerk all-renewable experiment. Biomass consists of plants and trees, which grow thanks to solar power. In growing, they extract carbon dioxide from the atmosphere. We will find out more about this in Part III. Biomass can be used to generate electricity directly. It can also be converted into biogas, which in turn can be used for electricity generation.

Most importantly, biogas can also be generated from many different types of biodegradable waste products. These can provide cheaper electricity than natural gas electricity generators as well as a more secure supply compared to imported gas.

There are a two important challenges facing electricity generation by biomass and biogas. First is the competition for land use with crops used for food, particularly in developing countries. This has been a major problem in the case of biomass grown for conversion to biofuels for transportation. The Renewable Energy Association maintains that there have been major improvements in sustainability standards in recent years. It is, of course, not a problem for biogas produced from waste.

The second major concern is about the carbon footprint of the entire biomass–biogas cycle. On the positive side, every carbon atom emitted into the atmosphere by burning biomass or biogas was pulled out of the atmosphere in the months before the crops were harvested. Hence these carbon atoms are recycled and, in principle, the whole cycle could be what is often called *carbon neutral*. In contrast, when burning fossil fuel, carbon atoms are released that were pulled from the atmosphere hundreds of millions of years ago. Fossil fuel carbon atoms are new arrivals in our atmosphere.

Any greenhouse gases emitted during the harvesting, transportation, conversion to fuel and waste disposal of biomass or biogas fuel will give that fuel a carbon footprint. In some cases, the carbon footprint could end up as high as that of natural gas. Later in this chapter, we will find that there are a number of biogas and biomass approaches to electricity generation with good carbon footprints. We will also meet approaches to generating biogas from waste in the next chapter that actually obviate a particularly nasty greenhouse gas.

As usual the Germans are ahead of the game. Their Federal Environment Agency proposes in their 'Energy target 2050: 100% renewable electricity supply' that the use of biomass for electricity generation (or even heat generation which we will discuss in the next chapter) is unnecessary. They believe that sufficient biogas can be produced from organic wastes and residues. They point out that it is more important that the biofuels from waste and residual biomass be used to replace fossil fuels in transportation.

In the next chapter, we will hear the views of a director of the first company to input biogas into a local gas grid. Their biomass is sustainably sourced with minimal extra carbon emission.

Energy efficiency

One more contribution to the solar revolution must never be forgotten. It is one of the most effective, low carbon, energy-generating approaches of them all: reducing electricity demand and natural gas demand by efficiency improvements.

Combined heat and power (CHP) is one of the finest large-scale examples of energy efficiency. It makes use of the heat that is released as a by-product whenever electricity is generated by whatever method. In terms of greenhouse gas reduction, CHP is best with a renewable electricity generator such as biogas or biomass, or geothermal as we met earlier. Denmark has pioneered CHP schemes, many co-operatively owned, in which hot water, heated by the waste energy from an electricity generator, passes in pipes through local heating networks. Originally the heat was the waste of fossil fuel powered electricity generators. Now many of these generators are being converted to biomass and biogas generation.

As with many of the solar technologies, there is both a large- and a small-scale version of energy efficiency. It is well known that thermal insulation and double glazing save on gas and electricity bills. In the UK you can get useful information on energy savings from the Energy Savings Trust.

I am concerned that successive UK governments have cut successful home insulation schemes and introduced replacements that turn out to be less popular, at least initially. In December 2013 the *Guardian* reported a 93 percent fall year on year in domestic loft installations when a new scheme was introduced by the coalition government.

In the next chapter, we will meet a number of new technologies that can replace domestic gas central heating, and hence dramatically reduce a home carbon footprint. The effectiveness of these new heating technologies is improved with good thermal insulation and double glazing. One clear example is the case of solar hot water where a large, well-insulated hot water tank will increase the effectiveness of a roof-solar thermal system. I am concerned that the switch to new low carbon technologies might be inhibited by requirements that the energy efficiency improvements are made first. If the new heating technology is really zero carbon in use it is only necessary to install the thermal insulation first if your priority is cost saving rather than carbon saving.

Before we finish with low carbon electricity generation and move on to low carbon gas and heating, we need to have some idea of how big the carbon savings are of the renewable electricity technologies.

Carbon footprints of solar and natural gas electricity

We now need to compare the carbon footprints of the different technologies that make up the solar cornucopia. I will also compare them with their main rivals for the attention of politicians: natural gas and nuclear power.

When comparing the carbon footprints of electricity-generating technologies, we need to take into account carbon dioxide emitted in all stages: construction, operation, production of any fuel, dismantling and waste disposal. Such a study is called a life cycle analysis (LCA).

LCAs are often presented in terms of carbon dioxide emissions. Remember that there are other gases that are responsible for global warming. Methane, another villain of our story, is a particularly dangerous greenhouse gas as we will see later. The best LCAs take all greenhouse gases into account and present *equivalent* carbon dioxide emissions.

The results of LCAs are usually given as the number of grams of carbon dioxide, or its equivalent, emitted for each kilowatt-hour (kWh) of electrical energy (gCO_2/kWh) emitted over a typical generator lifetime.

The UK government's Committee on Climate Change (CCC) has recommended that by 2030 electricity generation in Britain should be from sources that emit less than fifty grams of carbon dioxide for each kilowatt-hour of electrical energy generated (50 gCO_2/kWh). I agree. I believe it is crucial for our civilisation that all electricity generation, worldwide, should have a carbon footprint below this figure by 2030.

In 2014 Daniel Nugent and Benjamin Sovacool published a paper in *Energy Policy* that reviewed critically all the published LCAs of the renewables. They focused in particular on the large number of calculations for wind and PV generation. Their conclusions were presented as averages of carbon dioxide emissions from the LCAs which passed their strict quality controls.

I am pleased to be able to report that all the renewable technologies came in below the CCC finishing line. In first place, with a carbon footprint one fifth of the CCC limit ($10 \, gCO_2/kWh$) was the one-time winner of the race for the first public electricity supply: hydropower. Extremely close, in second place, came a technology we will talk about a lot: biogas electricity from anaerobic digestion. Nugent and Sovacool also reported results for seven different types of electricity generation from biomass. Despite the controversy discussed earlier, all came in below the CCC limit.

My own favourite, PV, crept in just under the CCC finishing line. There are reasons for this – let me explain. Remember that Cinderella's sisters were dressed in the most expensive crystalline silicon at the satellite ball? Crystalline silicon is expensive because a lot of electricity is needed to purify the relatively thick layers of silicon. In the LCAs studied most of this electricity came from high carbon sources. If you want crystalline silicon PV and you want it low carbon, choose silicon sourced in Norway. The electricity there is nearly all renewable. Alternatively, buy it soon. Most of the greenhouse gases will already have been emitted in manufacture. By 2030 the system will still be generating electrical power and it will be doing so with minuscule carbon emissions. The cheaper polycrystalline silicon finished more convincingly below the CCC line, as did the new thin-film, second generation technologies.

I conclude from the Nugent-Sovacool study that all the solar technologies have carbon footprints below the CCC limit. The simplest way to introduce a moratorium on all electricity generation apart from the renewables, as advocated in the last chapter, would be to insist on the $50 \, gCO_2/kWh$ limit for all new electricity generators here and now. As new electricity generation technologies are expected to have 20 year lifetimes, any country serious about hitting the CCC target by 2030 would need to impose the limit soon.

Where did electricity generation from natural gas finish in the Nugent-Sovacool stakes? *Nine times* higher that the CCC limit and 44 times higher than hydropower and anaerobic digestion. Natural gas – we will deal with you later!

The British coalition government is ignoring the advice of its own CCC. It has made clear it does not want to set any carbon limit before 2015. Even then, the limit they have been talking about is *nine times higher* than the CCC limit. Can you see why they want to set it nine times higher? This is to allow natural gas fired generators to be constructed. We will return to the politics of the CCC target in Chapter 10.

Where stands nuclear power? That question will require a section on its own.

The carbon footprint of nuclear electricity

There have been over two hundred papers on the carbon footprint of nuclear power in scientific journals in recent years. Three papers have critically reviewed the literature in the way Nugent and Sovacool compared renewable LCAs. The first was by Benjamin Sovacool himself. He reviewed 103 published LCA studies and whittled them down to 19, which had a reasonably rigorous scientific approach. The carbon footprints ranged from 3 to 200 gCO_2/kWh. The average carbon footprint was 66 gCO_2/kWh. This is above the CCC limit.

Jef Beerten and his collaborators took a different approach. They decided that such a wide range of results suggests that taking an average is not appropriate. So they studied closely just three typical published LCAs, one low, one in the middle and one high. Their extensive re-analysis did not bring them much closer together. The LCAs were 8, 58 and 117 gCO_2/kWh respectively.

In 2012, four years after Sovacool's paper, Ethan Warner and Garvin Heath found 274 papers containing nuclear LCAs. They filtered them down to 27 for further analysis. These yielded 99 estimates of carbon footprints which the authors describe as 'independent'. I do not see how the 99 estimates can be independent if they come from only 27 papers. I assume they mean 'distinct assumptions'. Their data for carbon emissions ranged up to 220 gCO_2/kWh. They did not quote an average value. Instead they reported that half the estimates were below 13 gCO_2/kWh, much lower than Sovacool's 66 gCO_2/kWh average. This is due to two LCAs that made 9 and 13 'distinct assumptions' that resulted in similar, very low carbon footprints. Had the 27 independent LCAs been averaged, the overall average would have been much closer to Sovacool's figure.

I must mention one unpublished, comparative study by Naser Odeh and colleagues from Ricardo-AEA because it was compiled for the CCC. AEA Technology was a spin-out from the UKAEA. Their report presents the spread of results from the Warner-Heath analysis but did not consider the reviews of Sovacool and Beerten et al. They focus on six other studies all with carbon footprints below 10 gCO_2/kWh. Only three of these LCAs are included in the Warner-Heath study.

Why the Ricardo-AEA team ignored all studies with higher carbon footprints is not clear. Since they based their own analysis on six LCAs all below 10 gCO_2/kWh, it is not surprising that they concluded that the carbon footprint of nuclear is around 6 gCO_2/kWh. In my view it would be unacceptable for a postgraduate student to ignore such a wide range of published data in his or her thesis. I wonder what the CCC made of this analysis.

I think it possible to explain why there is such a wide range of carbon footprints for nuclear. There are five important phases of the nuclear power life cycle. These are: reactor construction, running the reactor, preparing the fuel, decommissioning the reactor and the long-term storage of the spent fuel.

It is not surprising that estimates of the last three of the five phases differ widely. There are major uncertainties associated with fuel preparation, which we will meet later. Furthermore, not many reactors have been fully decommissioned and it is still not clear how or where spent fuel will finally be stored.

Beerten and co-workers concluded that widely varying assumptions about these problematic contributions were major factors in the differences between the three LCAs that they studied. Some LCAs included in the analyses of Sovacool and Warner-Heath did not provide an estimate for at least one of these three problematic phases. Uncertainties in these contributions lead to the wide range of carbon emissions. Ignoring one or more of them favours lower estimates.

Note that these three problematic contributions – fuel preparation, reactor decommissioning and waste fuel management – *either do not exist or are small for solar technologies*. Only biogas and biomass have any fuel or waste to worry about.

Now let's look at the carbon emissions in the contribution which nuclear and the renewables share: the construction phase. All the LCAs in the published reviews refer to electricity generators which have already been built. If the UK is to undercut the CCC limit in 2030 it will be the carbon footprint of future nuclear reactors that will matter. The two prototypes for the European Power Reactor (EPR) are still being built and are well behind schedule. How can one estimate their carbon emissions during construction if they are yet to be completed?

As of January 2014 the construction cost of one EPR at Hinkley Point C was five times higher than a hydropower dam (or a number of dams) providing the same electrical power. This higher price suggests higher carbon emissions. The EPR price reflects the high cost of more sophisticated engineering, manufacturing and transporting a steel pressure vessel, expensive high precision nuclear components and a steam generator. The only common cost with hydropower is the turbines. These will have the same power, though for nuclear they will use steam rather than water.

Many of these costs result from burning fossil fuels directly or in the generation of electricity. Construction of the EPR could result in up to five times higher carbon emissions for the same power. Assuming the reactor and the dams have the same lifetimes, and generate for the same time each year, the carbon footprint in terms of grams of CO_2 for each kWh of energy could be five times higher for the EPR than the dam. My estimate is only for the carbon dioxide emitted during construction. It ignores the costs of the problematic three nuclear contributions that are zero for the dam, so if anything it should be higher.

An LCA specifically for the EPR was one of the six new analyses below 10 gCO_2/kWh introduced in the Ricardo-AEA report. This EPR analysis claims the exact opposite to my estimate. The EPR carbon footprint is 2 gCO_2/kWh or one-fifth of the hydropower LCA. To me this analysis does not seem realistic.

Finally, let's look in more detail at one of the three problematic costs which most renewables do not have: the fuel cost. Nuclear fuel preparation begins with the mining of uranium-containing ore, crushing the ore and extracting the uranium from the powdered ore chemically. All three stages take a lot of energy, most of which comes from fossil fuels. The inescapable fact is that the lower the concentration of uranium in the ore, the higher the fossil fuel energy required to extract the uranium. This important point was first made to me by the Dutch physicist Storm van Leeuwen. We were both consultants to the Oxford Research Group for a period.

It is not difficult to understand why extracting uranium from low concentration ores will involve higher carbon emissions. Just think about prospectors panning for gold in streams in the Californian foothills. They were lucky in that the mining and crushing of the gold ore had already been done by the action of the mountain streams over millennia. The final stage, the extraction of the gold was done by laboriously sieving the sediment in the bed of the stream. The less concentrated the gold in the sediment, the longer the time and the more energy the prospectors expended sieving the sediment to find the same amount of gold. It is the same for uranium extraction. The only difference is that for 'longer time and more energy sieving' in the case of gold read 'more fossil-fuel energy expended' in the case of the mining, crushing and the chemical extraction of uranium.

Storm points out that the bulk of the ores in the World Nuclear Association data on uranium supply have uranium concentrations differing by a factor of about 20. It is an unavoidable consequence of the above argument that the energy input and therefore the carbon dioxide emitted in the extraction of the uranium will be at least 20 times as high for the ore with low uranium concentration as for ore with high concentration. To my mind, any estimate for the carbon footprint of nuclear power that doesn't show a strong increase in carbon emissions with falling uranium concentration in the ore is suspect.

The fact that van Leeuwen takes this dependence into account suggests to me that his calculations, which are at the high end of the values we have talked about, are probably more realistic than most.

I believe that when one or other of the EPR prototypes finally operates, and before serious construction work at Hinkley Point C starts, a full and complete LCA of the EPR should be produced by the manufacturers Areva.

The LCA should be peer reviewed by independent academics involved in the published LCAs. It should separately calculate the carbon emissions and their uncertainties from all phases of the life cycle. The assumptions made on greenhouse gas releases during mining, decommissioning, waste storage and environmental clean-up should be clearly explained. The effect of the uranium ore concentration should also be evaluated.

The solar technologies are all below the CCC limit and could supply all our electricity generation. Could they also replace natural gas, with its high carbon footprint, in heating our homes?

NINE

How Can We Reduce Our Carbon Emissions?

Give us the tools and we will finish the job.
Winston Churchill (1941)

I hope the last two chapters have convinced you that there are many renewable energy tools available that can dramatically reduce the carbon emissions from electricity generation, swiftly, sustainably and cheaply. Some of the tools originated in the USA. Churchill's wartime appeal was directed westward.

In the first half of this chapter, we will look at ways you can ensure your electricity supply at home is renewably generated. In the second half, we will meet some solar tools that can reduce the amount of natural gas burnt in your home and on a national scale.

How much would it cost to put PV on my roof?

We have met many examples of how power is more important than energy for PV. When a householder, or an electricity-generating company, buys a PV system, what matters is the price for a certain amount of power. So you see PV prices expressed in US dollars per watt ($/W) or UK pounds per watt (£/W).

What is happening to the price of solar panels? You will have seen between 10 and 16 of these rectangular shaped panels on a typical solar-powered house. For the five years before January 2009, the price for a first-generation PV panel remained quite steady despite rising demand, because of the silicon shortage described in Chapter 6. During those five years, the increasing construction cost of a new nuclear reactor (also expressed in dollars per watt) crossed over the steady cost of a PV panel. The cost of a new nuclear reactor has continued on this upward path, as described in the Bibliography.

January 2009 was a tipping point for the solar revolution. The German feed-in-tariff had at last achieved its aim. A market had been stimulated which led to mass production and falling prices. The market also stimulated new silicon supplies and new thin film cell manufacturers. The bad news for Western solar companies was that the Chinese were getting in on the act. They had built their own supply lines and massive one-gigawatt-a-year factories. By 2010, half the world's solar modules were being manufactured in China.

From this tipping point there was a precipitous fall in the price of PV panels. It fell from its January 2009 price (4 $/W) to *one-quarter* of that price (1 $/W) by June 2012. On the way down, the PV panel price dropped below two important renew-able costs, which are falling less steeply with time. The first was the cost of onshore wind power. The second was what is known as the *balance-of-systems* cost for new PV insulations. This is the cost of the scaffolding work, installation and the electronics for connection to the grid. Such costs now dominate over the module cost in the price of a new PV installation.

The PV price has now steadied, the feed-in-tariffs are being cut and the installation costs are not falling as fast as the panel price. So if you are thinking of investing in a PV system, don't delay in the hope that prices will continue to fall fast. I suggest you type 'solar panel price comparison' into your search engine. In 2013, while I was researching a system for our house in Frome, Somerset, the site Comparemysolar.co.uk carried price quotations from a number of UK solar PV suppliers. For a 3.5 kW, 14-panel system, which is on the large size for the average UK house, the prices range between £4,950 and £6,870 including installation. This website will tell you what feed-in-tariff you can expect. It will also work out how much electrical energy a system will deliver in a year.

Bear in mind that you will not be reducing your electricity bill by the full energy output of the system. You will have to buy back from the grid when the sun doesn't shine. There is uncertainty over how much you will save on your electricity bill in the UK: the Energy Savings Trust says 25 per cent, the Comparemysolar site assumes 50 per cent. But, given that the sun only shines for around 10 per cent of the year in the UK, the range 25 per cent to 50 per cent shows that in practice we use electricity more during the day when the sun shines.

The 50 per cent to 75 per cent of your PV energy you send to the grid should give you two glows of satisfaction: a financial one thinking about the FIT receipts (which in the UK are mainly based on how much electrical energy you generate rather than how much you send back) and an altruistic one from the knowledge your energy is going to the grid when it is most needed.

You can save more on your electricity bill with PV if you take simple measures: a sunny day is a good day for drying clothes on a line. With PV on the roof, it is clearly also the day to run your washing machine. I also described a new electronics box in Chapter 7 which will switch any excess electricity you generate into the immersion heater in your hot water tank. This will save on your hot water bill. Details are in the Bibliography.

Your roof is not particularly sunny or you live in a flat? I suggest you read on.

My roof is not suitable for PV

If your roof does not appear to have an ideal elevation, I suggest you make some calculations with the software on the

HOW CAN WE REDUCE OUR CARBON EMISSIONS?

Comparemysolar or similar sites. Work out the electrical energy you will get in a year on other elevations on your roof. On our house in Somerset, the elevations face approximately east–west. We recently decided to put 2 kW of PV panels east facing and 2 kW west facing. The calculations suggest we will generate around the same amount of electrical energy in a year as a 4 kW system with the ideal south-facing elevation. We do, however, expect bigger savings on our electricity bill because we will use more electric power at times when our system is generating. For example, on a sunny September morning at 9.00 a.m., when a south-facing system would not produce much power, our east-facing panels were enough to run the washing machine and we were trickling excess PV power into the immersion heater in the hot water tank.

The ideal system for a partially shaded roof or the vertical walls of a flat would be based on the 42.5 per cent efficient CPV cells. They could take up about half the area to produce the same amount of power. Sadly, the concentrator systems in which they need to be deployed are some way from implementation. My old group at Imperial is working on a system that would be ideal for sunny vertical walls as they fit into deep, double-glazed windows. We call these *smart* windows. They have transparent, sun-tracking blinds that are actually lenses, which focus the direct sunlight onto the high efficiency solar cells to generate electricity. On sunny days, the blinds cut the direct sunlight, reduce the air temperature in the building and hence reduce air conditioning demand. Being transparent, the blinds allow some diffuse (non-direct) sunlight into the building, thus providing natural, internal illumination without glare. In this way, a major problem with conventional blinds is avoided. If opaque blinds work properly, we turn the lights on when the sun shines.

As well as *generating* electricity, therefore, smart window systems *reduce* electricity demand by cutting the need for air conditioning and lighting. If the cells are cooled by water, the heated water can also be used in the building. Such systems would be ideal for flat dwellers or householders with walls facing east, south or west. They would be particularly useful in northern latitudes, where the sun never gets very high in the sky.

Sadly this technology has been difficult to commercialise. So for flat dwellers and those with partially shaded roofs who wish to lower their carbon footprint here and now, the solar thermal systems we met in Chapter 8 are a better buy. They occupy much less roof area than PV, can be mounted on vertical walls and should qualify for a subsidy. But do read on to the air source heat pump section before making your decision.

Flat dwellers could also try co-operating with neighbours or their housing association on a rooftop PV system or, if the building is high enough, a vertical-axis wind turbine.

One action that you can implement immediately to cut your carbon footprint, even if you have no savings at all, is described in the next section.

Cutting the carbon footprint of your electricity supply here and now

A few clicks on a mouse and possibly the odd telephone conversation are all that it takes to become a solar revolutionary. This is true in Britain even though it lags well behind in the solar revolution.

All you have to do is to type 'renewable electricity comparison' into your search engine. In the UK, this will bring up a number of sites that compare the electricity companies which supply only renewably generated electricity. Do look at the detail of their fuel mix. I used the Green Electricity Marketplace to decide that Good Energy were the company for me. One reason was that they really are 100 per cent renewable. Good Energy claim that it takes less than five minutes to switch, and they will contact your old supplier for you.

I am extremely happy with my Good Energy electricity supply.

I have been with them for two years and our lights have never gone out. That is what solar sceptics claim will happen with an all-renewable electricity supply.

Good Energies recently passed the 100,000 customer level milestone. They also came top of all British electricity suppliers in a customer satisfaction survey organised by the consumer magazine *Which?*

I had an informative discussion with a Good Energy spokesman Ed Gill. He confirmed that they have a number of large companies as customers. Hence another myth of the solar sceptics, that an all-solar electricity supply may be OK for domestic use but industry would grind to a halt, is disproved in practice.

Ed also confirmed the company had no intention of taking nuclear electricity. He outlined encouraging plans for the expansion of their own renewable generating capacity. He agreed that in the longer term their electric power should get cheaper than their competitors as the cost of installing renewable generators is falling and they don't have to worry about fuel prices. Sustainable bio-generators are only a very small proportion of their supply (0.6 per cent).

Good Energy have a tariff with cheaper night-time electricity. This is great for those with storage heaters. The only downside that I have found is that people with pre-payment meters cannot switch suppliers. This should be changed as I suspect it disproportionately affects those in fuel poverty.

Finally, Ed Gill confirmed that Good Energy would not be embarrassed should publication of *The Burning Answer* result in a surge of new customers. There are other green companies you might like to choose. So no excuses – get switching!

Cutting our use of natural gas

We now have the tools to make major cuts in the carbon footprint of our electricity generation. So now let's look at technologies that can dramatically reduce the amount of natural gas we burn.

THE HERE AND NOW OF THE SOLAR REVOLUTION

Remember that every atom of carbon emitted as carbon dioxide in the burning of natural gas had previously been safely locked up underground for hundreds of millions of years. Furthermore, the exploration, extraction, purification and transport of natural gas also generates large amounts of carbon dioxide and other greenhouse gases.

On the other hand, every carbon atom emitted by burning biogas generated from biomass crops was pulled out of the atmosphere in the months before the crops were harvested. Hence the carbon is recycled; it is not new carbon. Of course, one must ensure that in the harvesting, preparation and transport of the biogas as few extra greenhouse gases as possible are released. Also, the biomass growth should be sustainable and not detract from the growth of crops for food. We will meet examples of biogas generation from biomass where these conditions are fulfilled.

We will also meet some very encouraging examples of biogas produced from various types of waste. In this case, not only is the production not competitive with the growth of food crops but the whole cycle is particularly low carbon. Much of the component of our waste that we call *biodegradable,* if left to rot in fields or in landfill, would finally end up as methane. Remember that methane is one of the villains of our story as it is a particularly potent greenhouse gas. So it makes perfect sense to carefully convert our biodegradable waste to methane and then pump it into the national gas grid. It is not only a synthetic and low carbon version of natural gas, which itself is mainly methane, it also reduces the amount of greenhouse gases emitted were the waste left to rot.

Before we look in more detail at this ideal scenario, let's see what we can do individually to reduce the amount of natural gas we burn at home.

Reducing natural gas burning at home

As soon as you have switched to an all-renewable electricity supplier there are plenty of ways to cut the amount of natural gas you burn at home and reduce your gas bill. One obvious option is to switch from cooking by gas to cooking by electricity.

The major reason for burning natural gas in most domestic residences is the central heating. Why not consider replacing all or part of your central heating system with modern efficient electric radiators? There is information in the Bibliography to get you started.

There is still a view around that heating your house with electricity is expensive unless you make use of cheap rate electricity overnight. For example, the UK Energy Saving Trust still makes the outdated assertion on their website that 'Electricity is the most expensive and carbon intensive heating fuel available.'

The second part of the statement is clearly misleading. If your electricity comes from an all-renewable supplier, it is the least carbon intensive heating fuel available.

The first part is also misleading, but the explanation is more subtle. In the UK, gas supply companies now charge by the kWh like electricity companies. If you compare bills it looks as though the price of a kWh of gas is around half that of a kWh of night-time electricity and a third that of daytime electricity. But the amount of heat energy you get from a kWh of gas will depend on how efficient your gas boiler is. A good modern boiler is around 90% efficient, but an old boiler could have efficiency as low as 50%. So heating with an inefficient boiler could turn out to be as expensive as night time electricity.

By contrast a kWh of electrical energy can be turned into a kWh of heat energy extremely efficiently in a modern electric radiator. The electrons give up nearly all their energy of motion when they bash into to the atoms in the element in a radiator. This sets the atoms jiggling faster, just like they do in the element of an electric kettle, so the temperature of the radiator rises.

Bear in mind also that with electricity you don't have to purchase a boiler, pump or water pipes. Electric radiators can be added one room at a time as finances permit.

Finally, in operation, electricity can end up as cheap as gas because electrical radiators are easier to control than gas central heating. Ease of control reduces the heating bill. For example, many gas central heating systems rely on one central thermostat, which is sensitive to the temperature in one part of the house. The electric systems can have individual controlled thermostats on every radiator. These can be centrally controlled. Some modern gas radiators have thermostats, but how often do we go round turning down the radiators in rooms not in use?

I recently bought a small electric heater that gives instant radiative warmth. I still get unnecessary guilty feelings when I switch it on. My inhibitions about using electricity for heating are so ingrained. These modern electric heaters are more efficient than the old ones. My new radiant heater warms me up much more, and far quicker, than the old wire-wound electric bars that I shivered over while studying undergraduate physics. As long as your electricity is from an all-solar supplier, turning on a modern electric fire should be a guilt-free experience. At 10 a.m. on a cold but sunny November morning we had both bars of this radiant heater on and our east-facing PV system was still trickling excess electric power to the hot water tank. This was heating by free electricity, and carbon free as well: Energy Savings Trust please note.

There are other technologies that can replace gas central heating. They all require less power than electric radiators and are therefore likely to be cheaper to run, though at the present time they are still expensive to install, as we will find in later sections.

Heat pumps

Like the solar technologies in Chapter 8, the quotation from Ecclesiastes is appropriate yet again for heat pumps. The Industrial Revolution took off in the nineteenth century because engineers developed steam engines. These used the heat energy at high temperature produced by burning coal to generate steam that pushed pistons that powered the Industrial Revolution.

The pioneers who developed the first steam engines were practical types, not particularly interested in the physics that made them work or any theoretical ideas that might make them more efficient. The first person to think seriously about such matters was Sadi Carnot, a French army engineer. He was born in Paris a few years after the French Revolution. His family members included a number of senior government figures.

Physicists have a simple picture of the heat energy in steam. It is the energy of motion of the water molecules flying around extremely fast at high temperature. Some of this heat energy can be turned into power. The molecules can bounce off a piston, forcing it to move. The piston drives the wheels of an engine.

The steam molecules will have lost some of their energy of motion in creating this power, so the molecules rush around less fast and the temperature of the steam falls. Carnot's contribution was to show how the amount of useful power depends on the initial and final temperatures of the water molecules.

Carnot died of cholera in 1832 and, as a precaution, his notes were burned: one single manuscript and a few notes survived. His work was rediscovered and confirmed in 1849 by Lord Kelvin. He was Professor of Natural Philosophy at Glasgow for 53 years in the second half of the nineteenth century. Lord Kelvin was born in Belfast, Northern Ireland. He was awarded his peerage for his work on the transatlantic telegraph.

Kelvin realised the importance of Carnot's ideas and recast them into a mathematical form known as the *second law of thermodynamics*. The law quantifies how much heat energy from a hot gas or a hot solid object can be turned into useful power. As Carnot had thought, this depends on the initial and final temperatures. Having given up their heat energy, the molecules in the gas fly around at much lower speeds and the atoms of the solid object jiggle around much less.

Until Kelvin formulated his second law it had been thought that heat energy could only flow from a hot object to a colder one. Kelvin's version of the second law suggests that heat energy can flow from somewhere at a low temperature to another place at a higher temperature. But it can only do so if power is provided to make heat energy flow in this unexpected direction.

Kelvin realised his version of the second law could be applied backwards. The law not only describes how a steam engine works but also that a *heat pump* works the opposite way round. A heat pump uses, rather than generates, power. This power drives heat energy from a low temperature object to a higher temperature object, reducing the temperature of the first and raising the temperature of the second. We will see how this works in some examples.

There is still a brand of refrigerator known as 'Kelvinator', named after the discoverer of the heat pump that makes them work. Electricity provides the power so that food inside is cooled while the kitchen gets very slightly hotter. You can feel the higher temperature air, usually behind the fridge, when the heat pump (often called the *compressor*) is running.

Ground source heat pumps

Ground source heating is the small-scale version of geothermal power, which we met in Chapter 8. It is already well established for heating all types of buildings in Scandinavia. As an alternative to natural gas for heating, it can reduce gas bills to zero, dramatically cut your carbon footprint and greatly reduce electricity bills for heating.

HOW CAN WE REDUCE OUR CARBON EMISSIONS?

One only needs to drill down around two metres below ground to find a region at a fairly constant temperature of 11 °C. At this distance below the earth's surface, the main contribution to the heat has come from sunlight absorbed at the surface of the earth as described in Chapter 7. Heat flows slowly through earth and rock as I described in Chapter 8. Even the first two metres of earth and stone below the earth's surface provide good thermal insulation. The temperature remains around 11 °C throughout the year – even in Sweden! The summer sunlight is being stored as heat energy for our use in the winter. This is another answer to those sceptics who claim you cannot store solar energy.

This solar heat energy could be used to heat a typical house. Electricity can pump water (with anti-freeze) round a continuous circuit of tubing laid two metres underground in the garden. If the system works well, the water should arrive in the house at a temperature approaching 11 °C. This heat energy could be used, in a device known as a *heat exchanger*, to warm the water that is pumped round your central heating circuit. However, this would not be very effective. Your house would not get anywhere near 11 °C and, as you know, that is not very warm. The real breakthrough was to replace the heat exchanger by a heat pump working in the way Lord Kelvin proposed.

So how does the heat pump work in a ground source heating system? It consists of a third closed circuit of tubing containing a refrigerant and a compressor, like the ones in your fridge. The tubing is wound so it can exchange heat with both the tube from underground on one side of the compressor and your central heating pipes on the other side. The heat energy from the underground circuit makes the refrigerant liquid evaporate to form a gas. This is just like water molecules in the sea evaporating to water vapour thanks to solar energy, which we met in Chapter 6.

Here comes the really clever part. Electricity is used to compress the refrigerant gas. Remember I said the higher the temperature of a gas, the higher the speed of the gas molecules? If you press down on the refrigerant gas molecules with a piston, you increase the speed of the gas molecules and hence the gas temperature. Think of the gas molecules bouncing off the incoming piston and rebounding faster. It is just like that satisfying moment when you make a good return volley playing tennis. If you apply power with your racket at the right instant, you can return the ball faster, and with higher energy of motion, than you received it. A heat pump compressor can increase the temperature of the vapour considerably; temperatures above 40 °C can be achieved efficiently. The temperature of the water in your central heating circuit therefore rises much higher than if the underground water was used in a heat exchanger.

The heat energy given up to the central heating circuit means the gas in the heat pump circuit cools. It condenses back to a liquid, ready to evaporate again on reaching the underground water.

It sounds too good to be true, using cold water to raise the temperature of water higher. Does a heat pump violate the principle of conservation of energy? Remember, physicists consider that principle so sacred that they call it the *first law of thermodynamics*. No, a heat pump is a good example of energy conservation as well as a good example of the second law of thermodynamics working backwards.

A good heat pump system can pump three kilowatt-hours (3 kWh) of heat energy from underground into the house as long as it is running at 1 kW of electric power for one hour (1 kWh). Thanks to the conservation of energy the house ends up with 4 kWh of heat energy. It is as if the house is being heated by four one-kilowatt electric fires on the electricity required for one, thanks to the solar energy stored underground.

The heat energy from under the garden lawn is free once the installation is paid for. It also comes carbon free apart from the carbon released in manufacture and installation. You have to pay for the electricity needed to compress the gas. But you only use a quarter of the electricity you would need if you produced all

4 kWh of heat energy by electric radiators so it should end up even cheaper than gas. If you get your electricity from an all-solar supplier (have you switched yet?), then this form of heating is completely carbon free in operation.

One of the great benefits of ground source heat pump technology is that some versions, which have *reversible* heat pumps, can be made to run backwards in the summer. A reversible heat pump will suck excess internal heat energy out of the building back underground. It is easier for the heat pump to do this as heat normally likes to go from hot to less hot.

In this way the ground source heating system is providing air conditioning in the summer. The ideal of course is to use solar panels on the roof to power this air-conditioning function of heat pumps. When do you need air conditioning?

I like the idea that a ground source heat pump running backwards in summer not only cools the building but replenishes the store of solar energy underground for use during the winter. This is yet another riposte to anyone claiming that you cannot store solar energy long term. The reversible heat pumps can take the surplus heat energy in your house in summer and add it to the underground store and then remove it again in winter.

The fact that Sweden has for many years been leading the way in exploiting this resource is a great advertisement for the practicality of heat pumps. It is also demonstrates that solar energy can be stored over winter even in the coldest climates. My colleague Jan-Gustav Werthen tells me that ground source heat pumps work well in Lapland, above the Arctic Circle. The Swedish experience of heat pumps has been so positive that for more than a decade 90 per cent of new houses in Sweden have been built with them installed. There is a lesson here for all European countries south of Sweden. They all have a higher solar resource.

I am very pleased to be able to report that ground source heat pump technology in the UK is spreading from domestic to commercial applications. A major UK supermarket chain that already has

PV on many of its buildings is now adding ground source heat pumps to tap the heating resource under its car parks and improve the efficiency of its refrigerators.

I have had a very stimulating discussion with Dmitriy Zaynulin, the founder and CTO of Greenfield Energy. His company have already installed systems that are working well in a number of branches. There are many exciting features of their novel approach.

First, they use oil-industry techniques to drill down to 200 metres underground. This is a lot further than the two metres in a domestic back-garden. They are exploiting geothermal energy rather than solar and have access to more heat energy at a higher temperature than the 11 °C under the lawn.

Very impressively, the heat energy and the temperature are sufficient to enable heat pumps at ground level to supply all the central heating and all the hot water that the store needs. They also have heat pumps which work like the reversible heat pump I described. They can provide air conditioning at any time it is required. They take the excess heat inside the supermarket and store it under the car park.

Finally, freezer cabinets are an important part of any super-market. Traditionally these are cooled by more powerful versions of the heat pump in your refrigerator. But Dmitriy explained to me that by linking the refrigeration units into the same circuit as the heat pumps, the two systems assist each other. The stable temperatures coming from underground help the refrigeration systems work more efficiently. Then, like using the reversible heat pump for air conditioning, the heat sucked out from the freezer cabinets can be stored under the car park and later recycled when-ever the store needs heating.

The whole impressive network of heat pumps and heating and cooling circuits is computer controlled to keep the shopping expe-rience pleasant, while the heat pumps hum away efficiently in the background. A fine example of Victorian technology made practical by the semiconductor revolution.

HOW CAN WE REDUCE OUR CARBON EMISSIONS?

Of course, there are plenty of people who would like to switch to zero carbon heating who haven't got a garden or carpark that can be dug up. Many also don't have the necessary £9,000 or more revealed by visiting the websites suggested in the Bibliography. What can be done for them? Amazingly, heat pumps can extract energy from the air on a cold day, as we shall see.

Air source heat pumps

Air source heat pumps work because, as far as the physics is concerned, there is nothing particularly different about air at 20 °C inside a house and air at 0 °C outside. True, the air molecules outside in the cold are rushing around with average energy of motion 7 per cent lower than the warmer molecules inside the house. However, the ones outside are still moving at high speed.

It is also true that at 0 °C water freezes to ice. But physicists picture atoms in the solid ice still jiggling around their average positions and so they still have a lot of heat energy. To reduce the energy of motion of the molecules in a gas to zero, or to stop the atoms in a solid jiggling around, you have to go to an extremely low temperature of −273 °C. My colleagues in solid state physics have to be very ingenious to get near that temperature in their laboratories. There is still plenty of usable heat energy in air molecules at −10 °C: a relatively high temperature to a well-wrapped-up physicist.

The air source heat pump unit doesn't need all the underground pipe-work of the ground source variety and is therefore cheaper. You can find prices starting at £6,000 for heating a detached house on the websites in the Bibliography. The cost for a flat would be less, particularly if you can get a subsidy.

The main problem is that an air source heat pump doesn't have the all-year, underground solar energy store to rely on. Though a good air source heat pump can raise the air temperature so as to supply heat energy that is four times the consumed electrical energy, just like a ground source system, it has to start from a lower outside air temperature in winter. Hence air source heat pumps are less effective in winter than ground source. Also, they work best in new-build houses specifically designed for their use. When retrofitted to older properties, it has sometimes proved difficult to achieve the performance claimed by manufacturers. It may be necessary to increase the size of radiators. Air source heat pumps work well supplying under-floor heating, but that is particularly expensive to retrofit.

Air source systems are mounted on external walls, so they are very suitable for flats. Some of them look something like the air-conditioning units that can be seen on the walls of commercial buildings with their distinctive, large circular grills. This resemblance is no coincidence. Just like ground source systems, reversible heat pumps can be fitted in air source systems so they can be run as air-conditioning units in the summer.

Recently, a number of systems have become available in the UK, described as solar 'thermodynamic', that combine a heat pump with a solar thermal collector. The manufacturers claim that a unit the size of a door (which can be mounted on a vertical wall) can supply all the hot water a house needs. The UK Energy Savings Trust cautioned in 2013 that these thermodynamic systems were still new and had yet to pass all the accreditation tests. The systems have the advantage for hot water, over solar thermal alone, that you are guaranteed near carbon-free hot water when you want it, any time, winter or summer. In winter, you will find you will be using small amounts of electricity to run the heat pump. But your electricity will be all-solar by then will it not?

Can I switch to a green gas supplier for my heating?

Sadly, many of these low carbon replacements for gas heating are relatively new, at least in Britain, and therefore still quite expensive, though often grants are available to lessen the pain. Wouldn't it be great if it were possible simply to switch one's gas supplier to an all-renewable gas supply just as one can switch to an all-solar electricity supplier?

Such an option may be closer than you think. The Renewable Energy Association Ltd (REAL) has initiated a Green Gas Certification scheme for companies who want to supply biogas to the national grid. Then, when there are enough suppliers, companies like Good Energy will be able to supply customers with biogas along with their all-renewable electricity.

The biogas replacements for natural gas are called by at least three names: green gas, substitute natural gas (SNG) or bio-methane. Natural gas itself consists mainly of methane. Biomethane could be the super-hero that vanquishes that particular villain.

I have had a very informative chat with Nick Finding, a director of J V Energen. In 2012 his company became the first in the UK to inject biomethane into a local gas distribution network. Nick anticipates that between 10 and 20 green gas suppliers could be coming on the grid by the end of 2014. Companies might then be able to start offering green gas to domestic customers.

J V Energen's plant, near Dorchester in the south-west of the UK, takes an interesting mixture of feedstocks from local sources. Partly this is maize silage. This is grown in the fallow season of the agricultural cycle and regenerates the soil. This use of biomass is less likely to compete with food crops for humans, though it does with feedstock for animals. The other main component is food waste.

Have a look at the video on the J V Energen site. It is very impressive how energy efficient the whole system is. In addition to injecting enough biomethane to heat over 3,000 homes into the local gas grid, some biogas is burned in their own CHP plant. This generates electricity used on site and exports enough to the national electricity grid for over 500 homes. The waste heat from the electricity generation is used for heating on site. So their system is another good example of a CHP plant. Finally, the waste from their biogas production is a very effective agricultural fertiliser.

If you visit the J V Energen website, you will find an explanation of the way their biogas is generated. It is produced by anaerobic digestion (AD) which, you may recall, finished a close second in the Nugent-Sovacool low carbon stakes. Air is excluded from the process, hence the name anaerobic; bacteria reduce the organic matter to biogas plus liquid and solid waste rather like our guts work, hence the name digestion.

AD can be used on many forms of organic waste. Its use on livestock waste is particularly sensible. Not only can it produce biomethane for the gas grid, power for the electricity grid and waste heat for local use, the fertiliser that is the waste of the digestion can replace the modern fertilisers that are fossil fuel based. Finally, and most importantly, AD actually *reduces* emissions of the villainous methane into the atmosphere. If the livestock waste had been left to rot naturally on the farm it would eventually have decayed into methane.

Clearly AD is a win–win renewable technology. Sadly, according to the Renewable Energy Association, the UK government is not supporting it strongly enough. I thought the Conservatives were the farmers' party? The REA also points out that 20 million Chinese households produce biogas for cooking by a low-tech version of AD. This is yet another example of the relevance of the Ecclesiastes epigram in the last chapter.

Another source of biomethane for the gas grid is known by the not very elegant name of *waste gasification*. Advanced Plasma Power call their version *Gasplasma*. They convert many types of waste including household waste that is not recycled and existing landfill to biomethane for injection into the gas grid. Their systems can be accommodated in a standard industrial warehouse with very low environmental emissions.

There is great potential for biomethane to replace large amounts of natural gas on national gas grids. The National Grid, the company responsible for the UK gas grid, believes that biomethane from all forms of waste could supply nearly half the UK domestic heating requirements by 2020.

If you are from the farming community and already send waste for AD, well done and welcome to the revolution! I hope you can see there is a great opportunity to reduce your carbon footprint and earn some useful revenue. Please pass on the good news to other farmers. The rest of us – the city and town dwellers – while awaiting the opportunity to opt for a green gas supplier, can join a local community renewable energy group and support them in producing biomethane. We will meet some encouraging examples of such groups in the next section.

Grass-roots support for the solar revolution

The number of local community-owned power systems in the UK is growing. In Denmark and Germany co-operatively owned schemes are nothing new under the sun. In 2012, 50 per cent of wind turbines in Germany and 90 per cent in Denmark were co-operatively owned. Feldheim, a village in Germany with 150 inhabitants, claims to be the world's first 100 per cent carbon-neutral community.

There is an important added benefit of involving the local community in renewable energy. There is evidence that there is much less local opposition to the installation of wind generators when the local community has benefitted. As PV farms and the domes of anaerobic digesters proliferate, one can anticipate local opposition from certain quarters will grow. That too can be deflected if it is clear there is a local benefit.

THE HERE AND NOW OF THE SOLAR REVOLUTION

Pure Leapfrog is a business-led charity that has supported over 100 community-run renewable projects in Britain. One case study describes how the thousand or so residents of Ashton Hayes in Cheshire aim to become 'England's first carbon-neutral community'. They calculate they have already reduced carbon emissions by 23 per cent. I particularly like their proposed renewably powered CHP plant, which will provide heat for a sports pavilion and electricity for their shared electric car scheme.

A second case study on the Pure Leapfrog site, which is also impressive, comes from an urban environment. Repowering London is based in Brixton in South London. They emphasise local ownership of renewable projects, fuel poverty reduction and employing local people to install their PV systems.

Yet another approach is taken by our local Bath and West Community Energy group. They have formed a Community Benefit Society, which has raised £1.25 million in two share offers to local residents. Their main installations so far have been PV, but I was particularly impressed by their proposed renovation of the wheel of an old watermill on the river Avon to generate electric power.

I hope you will act on some of the suggestions in this chapter. You can greatly reduce your carbon footprint, and that of your local community, with tools that are available here and now. These all relate to the carbon released in electric power generation and in heating. This is the main contribution to our carbon emissions. The second biggest contribution comes from transport related carbon releases. Before we consider ways to reduce these, we should try to answer the responses that governments are likely to make to a call for a moratorium on building new electricity generators apart from renewable ones.

TEN

The Politics of the Solar Revolution

Science is one thing, wisdom is another.

Arthur Eddington

This chapter will consider two burning questions that have been smouldering throughout *The Burning Answer*. Why is the solar revolution developing at very different speeds in different countries? What can we do to speed up the revolution in the countries lagging behind?

We have seen that Britain is way behind Germany in introducing solar technologies, so I will focus mainly on these two countries and the political and technical factors that have given Germany its lead. I will also answer the arguments made by the British government, its scientific advisers and the scientific establishment as to why an all-renewable electricity supply is not appropriate for the UK. Finally, you'll see how we can start to counter the influence of the fossil fuel lobby, which will certainly oppose any call for an all-renewable energy supply.

We begin with Germany and the political decisions that started it on its path to the leadership of the solar revolution.

The Green Party legacy

The political origins of Germany's remarkable love affair with renewable energy began in the last years of the twentieth century when the Green Party was governing in coalition with the Social Democrats. The coalition government made two important decisions, to subsidise a 100,000-roof PV programme and not to replace nuclear reactors as they became obsolete. The 100,000-roof programme was a great success, so the coalition initiated the feed-in-tariff (FIT) policy. As we saw in Chapter 7, this too has been extremely successful. A steady exponential rise in PV installations was maintained in Germany for more than a decade and a half. It stimulated the formation of a market, which led to mass production and a dramatic fall in the price of PV panels worldwide.

There are a number of reasons why the German FIT has been so popular that successive governments of different political complexions have supported it. First, as we saw in Chapter 7, the price paid for the electrical energy PV generators feed into the grid was set to fall with time in a way which everyone was aware of; hence investments could be planned with confidence. Second, many of the renewable power systems are community owned. Third, the FIT is, strictly speaking, not a subsidy paid for out of general taxation like the fossil fuel subsidies we will meet later in this chapter. It is a levy paid by an increase in the cost of the electrical energy supplied to most, but not all, German electrical consumers. Indeed, many German industries that are large electricity consumers do not pay the FIT.

Much of German industry is benefitting from the fall in the peak price of electricity described in Chapter 7 thanks to a levy that they don't have to pay. Germany is also benefitting from the extra jobs and economic activity associated with new PV installations. These could be some of the reasons why Germany is climbing out of the 2008 recession faster than the rest of Europe.

The commitment of successive German governments to PV, wind and biogas electricity generation probably owes much to the German Red–Green coalition decision not to replace nuclear reactors as they reached the end of their life. In the light of the Fukushima disaster, this decision was confirmed. The die had already been set before Fukushima. In 2010, emboldened I am sure by the results of the Kombikraftwerk experiment that was funded by the German Economics ministry, the German Federal Environmental Agency set a target date of 2050 by which time Germany will have a 100 per cent renewable electricity supply.

Which country will be first to an all-solar electricity supply?

The front-runners in the race to achieve an all-renewable electricity supply are clear: both Iceland and Norway currently produce around 99 per cent of their electrical energy from renewable sources. That is mainly because they both have a large hydropower contribution. As I discussed in Chapter 8, Iceland also has plenty of easily exploitable geothermal energy.

Are there candidates among the European countries not blessed by massive amounts of hydropower who might beat Germany to an all-solar electricity supply? In February 2011, the Danish government announced the target of 2050 for all their energy use, not just electricity, to be renewably generated. They also announced that renewables would supply more than 60 per cent of electricity production in 2020. As the pioneers of wind power they might well beat Germany to an all-solar electricity supply. They have also pioneered local, community-owned, combined heat and power schemes, which are increasingly being powered by biogas.

An interesting challenge to Denmark and Germany is emerging.

Scotland has announced a target date of 2020 for 100 per cent electricity demand being supplied by renewable sources. The contrast with the Westminster parliament could hardly be greater, as we will now see.

The UK government position

At the time of writing, the British coalition government has refused to enshrine into law the recommendation of its own Committee on Climate Change (CCC) that by 2030 electricity generation should have a low carbon footprint (less than 50 gCO_2/kWh). We found in Chapter 8 that all the solar technologies are below that limit. As new electricity generators are expected to have a lifetime greater than 20 years, the only way to ensure the CCC limit is achieved by 2030 is to impose that limit as soon as possible. I think it therefore would be hard for the CCC to oppose our call for all new electricity generation to be renewable.

Some reasons why the UK government believes an all-solar electricity supply is impractical can be found in an interview given to *The Sunday Times* by the chief scientific adviser to the UK Department of Energy and Climate Change (DECC), Professor David MacKay. We met MacKay's book, *Sustainable Energy – Without the Hot Air*, in Chapter 8.

Jonathan Leake, the *Sunday Times*' environment editor interviewed MacKay and filed a story, which appeared on 20 November 2011 under the headline 'Green energy could blot out countryside'.

MacKay told *The Sunday Times* that 'if a country opted for solar power to dominate its energy supply it would need to cover a third of the land in solar panels'. It will be no surprise that I disagree strongly with this statement. I believe that if the sunny elevations of all roofs in the UK were covered with 13 per cent efficient panels, the PV would generate as much electrical energy as is consumed inside these buildings. Hence *no new land at all*, only the roofs of existing buildings, need be covered by PV cells to supply the two-thirds of the UK electrical energy which is consumed in those buildings.

Can an estimate of 'a third of the land' be reconciled with 'no new land at all'? I will show you how. Please bear with me.

In the *Sunday Times* article, MacKay said of the renewables 'they all deliver about 2.5 W per square metre'. That may be true for some of the renewables, but PV can deliver a lot more power. A photograph in *Sustainable Energy – Without the Hot Air* shows a PV array on a house in Cambridgeshire that generates 20 W per square metre of panel. I assume MacKay was thinking that the countryside would be covered by solar farms like the German one in his book, which generates 5 W per square metre.

These are average powers that MacKay quotes in his book. You can find there that the 2.5 W per square metre in *The Sunday Times* corresponds to PV panels of 2.5 per cent efficiency. My calculation assumes 13 per cent efficient panels close together on rooftops rather than panels spaced apart in fields. Hence the area required in MacKay's calculation is 5.2 times larger than mine (13/2.5).

The Sunday Times goes on to quote MacKay as saying that in Britain energy consumption is 'about 125 kWh per day per person'. This number is calculated some way through his book. It turns out to be an estimate of the *total* amount of all types of energy consumption, not just electrical energy. To be fair to MacKay, the words 'electricity' and 'electrical energy' do not appear anywhere in the *Sunday Times* article. However, I am sure that many *Sunday Times* readers would have concluded that he was talking about electricity as he was comparing the renewables with nuclear power.

How much of MacKay's figure for UK energy requirement of 125 kWh per day per person is due to electrical energy? In his book, Professor MacKay estimates the electricity contribution at 18 kWh per day per person. If we wanted to calculate the area of

UK land required to satisfy just the UK *electrical* energy demand, the area calculated by MacKay is 6.9 times too large (125/18).

To find the area of the British countryside that needs to be covered by 13 per cent PV cells so as to 'dominate' the electricity supply of the UK, one must divide MacKay's estimate of 1/3 of the country by 5.2 for the efficiency underestimate and by 6.9 because he included many other forms of energy consumption apart from electricity. That means MacKay's calculation of 1/3 of the country (33.3 per cent) is nearly *36 times* higher than my estimate. Dividing his estimate by this number (5.2 × 6.9) gives just under 1 per cent as the area of Britain that must be covered by 13 per cent PV cells to generate all the country's electrical energy.

It was a moment of some satisfaction when I exchanged my own assumptions for MacKay's figure in this way and turned his 1/3 of the land into a number around 1 per cent of the area of the UK. What did this remind me of? Just over 1 per cent is the land area of Britain taken up by buildings according to data in MacKay's book.

As long as you adjust MacKay's figure for the efficiency of PV on rooftops rather than an average efficiency in open fields and for *electrical* energy consumption rather than all energy, his calculation and mine can be reconciled. My own view, that all the electrical energy consumed in buildings in the UK can be supplied by covering the roofs with 13 per cent efficient solar cells, is consistent with his figure of 1/3 of the countryside. Both calculations are 'correct'. It is the assumptions that are different.

Another critic of PV is the environmental journalist George Monbiot. He strongly opposed the introduction of the UK feed-in-tariff in 2010. Some reasons for Monbiot's opposition to PV can be found in his book *Heat: How to Stop the Planet Burning*. There he criticises the statement by Jeremy Leggett, the solar revolutionary who founded Solarcentury, that 'Even in the cloudy UK, more electricity than the nation currently uses could be generated by putting PV roof tiles on all suitable roofs.' Monbiot's view is that 'there isn't enough sun and it shines at the wrong time'. I disagree on both counts. Chapter 7 showed that the PV *power* contribution peaks at the right time: close to the maximum daytime electricity demand. There is also enough sun on rooftops to supply all the electrical *energy* consumed inside. Even Professor MacKay, another solar sceptic, has made calculations that, for the same assumptions as Leggett, are consistent with Leggett's statement and with my own.

Another example of Professor MacKay's scepticism about renewable energy can be found in a memorandum he wrote for the House of Commons, Environmental Audit Committee in 2009. You can find the reference in the Bibliography. It was written in response to the question, 'Is it technically possible to decarbonise Britain by 2050?'

MacKay predicts that by 2050 the UK will need a massive 70 GW of nuclear power, as large as France has currently. In contrast, MacKay's prediction for PV is that expanding 'at roughly the maximum rate I think is plausibly achievable' there will be 7.5 GW of PV in the UK by 2050.

Had MacKay looked to Germany for guidance on the expansion of PV as he looked to France for nuclear, his prediction would have been very different. It took Germany just over 10 years to expand PV installations from where Britain was in 2008 to 7.5 GW. So if PV in Britain expands at the same rate as Germany achieved we could reach 7.5 GW by 2019. The German experience suggests that the UK could have as much PV as MacKay predicts for 2050 at the 'maximum rate ... achievable', but *31 years* earlier.

A final example of the way nuclear is favoured over the renewables in UK government circles is found in a computer program produced in Professor MacKay's department. This deserves a whole section to itself.

The program

The UK Department of Energy and Climate Change (DECC) has produced a computer program 'Pathways to 2050' so that members of the public can explore their own strategies to reduce the UK carbon emissions to 80 per cent of the 1990 levels by 2050. It is a great idea that the public can participate in exploring possible scenarios. The problem is the program strongly favours the nuclear option over the renewables.

The very basis of the program favours nuclear over solar because it works in terms of energy flows. As the DECC report '2050 Pathways Analysis' puts it, the program 'allows people to explore the combinations of effort which meet the emissions target while matching energy supply and demand'. You now know from the Kombikraftwerk experiment that to satisfy electricity demand it is necessary to match the different *power* supplies not the *energy*.

Using energy flows rather than power means the DECC program does not take into account the advantage that peak PV power coincides closely with peak daytime demand. Remember that this meant that, with only 3 per cent of PV electrical *energy* on the grid, the peak wholesale price of electricity in Germany fell below the night-time price in June 2011. This boost for German industry could not be predicted by DECC's program.

Energy flow calculations favour nuclear electricity over PV by around seven times. They also favour nuclear over wind by two to three times. This is because the sun only shines around 10 per cent of the year in the UK and the winds only blow strongly enough for around one-third of a year. A nuclear reactor with the same nominal power as a number of PV farms will produce seven times as much electrical energy in a year. But such a calculation will not tell you if that electricity comes at times when it is needed or not.

'Pathways to 2050' favours the nuclear option in other ways. The DECC calculations assume that the next generation of nuclear reactors will produce electricity for 80 per cent of the year, though none of them have worked as yet. In practice, UK nuclear reactors have struggled to generate electricity between 50 to 70 per cent of the time over recent years.

There is one part of the program where power rather than energy does feature. There is a 'stress test' of how the chosen options cope with adverse weather conditions, such as very little wind power for several days. This again favours nuclear over the renewables. There is no stress test for the sudden loss of power from a massive 1.6 GW nuclear power station. This would happen in a power cut on safety grounds. I believe the best 'stress test' of how a national grid can cope with adverse weather conditions is the real-time balancing of renewable power supply with power demand continuously over the year as in the Kombikraftwerk experiment.

The evidence suggests that the DECC calculations favour nuclear over the renewables. This supports the political decision, which had already been made in 2006, to build new nuclear reactors. I was therefore particularly impressed when two colleagues in Pugwash, David Elliott and David Finney, were able to work their way through the options in 'Pathways to 2050' to demonstrate a 'High Renewables Pathway'. This achieved the 80 per cent carbon reduction by 2050 without any new nuclear reactors. They would have achieved 100 per cent had the program allowed them to store excess wind energy. Their work is further evidence confirming one important message of this book: an all-renewable electricity supply is achievable, quickly, cheaply and safely. What is missing is the political will.

The national laboratories

In Chapter 4 we met a number of possible reasons why successive British governments have been fixated on reviving nuclear power. Later in this chapter, we will consider the influence of the pro-nuclear lobby on British governments and the fossil fuel lobby on all governments. First, though, what influence do scientists have on the solar revolution?

The impressive expansion of PV and wind power installations in Germany has been helped by the support of at least two national laboratories: the Fraunhofer ISE photovoltaic laboratory in Freiberg and the Fraunhofer IWES wind power laboratory in Kassel. The latter now looks after the ongoing Kombikraftwerk project that I described in Chapter 7. All of German industry has benefitted from the excellence of the 67 Fraunhofer applied science laboratories with their 23,000 staff who are mainly scientists and engineers. They have all been set up, since the end of the Second World War, by the non-profit society Fraunhofer-Gesellschaft, with national and state government support. The ISE and IWES laboratories have made particular impact, because the industries they support are based on new technologies.

The excellent contribution made by these specialist Fraunhofer laboratories to German renewable energy achievements is reinforced by comparison with the US situation. In 2011, the US had one-sixth of the PV installations of Germany 2011 despite being a larger country and a world leader in semiconductor research. The US has had a National Renewable Energy Laboratory (NREL) since President Carter's day but, as the name implies, work is split over all the solar technologies. Having visited NREL a number of times, I have been impressed by the laboratory's contributions despite having a budget far smaller than the US national labs I worked in during my particle physics days. The US Department of Energy's 2013 fiscal year budget for NREL is below that of either the Fermi or Stanford particle physics laboratories. The three national particle and nuclear laboratories – Argonne, Berkeley and Brookhaven – have more than twice the budget of NREL. The three big defence laboratories at Livermore, Los Alamos and Sandia each had a 2013 budget between five and eight times that of NREL. These figures, taken from the reference in the

Bibliography, represent the US Department of Energy support. Any US Department of Defense support is on top of that and will, of course, favour the bigger laboratories. Sadly, renewable energy is still a Cinderella research activity in the US. This goes some way to explaining why it lags so far behind Germany in PV installations.

One reason Fraunhofer ISE and NREL have taken the lead internationally in photovoltaic research and development is that there are few national PV laboratories in other countries with anything like the stature or funding of these two. The contrast with particle physics could hardly be stronger. There is a world-leading, quite literally, international laboratory at CERN in Geneva and most developed nations have their own national particle physics laboratory supporting university research.

The UK has no dedicated national PV laboratory. The National Renewable Energy Centre (Narec) in Northumbria has a PV Technology Centre. I am aware that its 15 full-time employees are doing good work and I wish them luck as they were privatised in 2012. However, it would be invidious to compare them with Fraunhofer ISE with its 510 permanent staff members or indeed the UK National Nuclear Laboratory with 550 research staff.

National laboratories are extremely important for emerging technologies and industries. I can provide an example from my own experience. If the UK had a dedicated PV laboratory like the Fraunhofer ISE, my company QuantaSol might still be providing green jobs in the UK. I have mentioned the quantum wells that were our unique technology. Early on, we used the wells to enhance the efficiency of a GaAs cell to beat a 21-year-old world record in concentrated sunlight, though sadly the record only lasted for six months. Then we set out to use quantum wells to enhance the efficiency of multi-junction cells.

We had the offer of a very significant amount of venture capital funding, but first we had to hit a tough milestone. We failed to do so, though the company later went well past the milestone. This was one of the most traumatic times of my life. The problem was not with the quantum wells, which were working well, but with the conventional part of the cell where contact was made to the outside world. The Fraunhofer ISE could have helped us with this part of the technology. However, we decided against asking them for a number of reasons. A UK national solar cell laboratory like the Fraunhofer ISE would have been able to help us to keep what turned out to be world-leading technology in the UK.

The scientific establishment

It is not only the absence of laboratories like the Fraunhofer Institutes that is holding back the development of renewables in the UK. I am disappointed to report that my experience of the scientific societies in the UK is that, despite their awareness of the consequences of global warming, in general they have not been particularly supportive of the renewable technologies.

My concerns about the Royal Society, for example, began when I read the opinion pieces in the national press written by Lord May of Oxford when he was President of the Royal Society in 2005. I accept these were his personal views rather than those of the society. His articles seemed very well focused when he drew attention to the activities of climate change deniers funded by the US oil industry. However, in his final speech as president he also accused NGOs opposed to nuclear power of the same fault of 'fundamentalism' and the mis-representation of scientific facts. Given that the Greenpeace Trust were the first funding body to support my solar cell research, you can see why I disagree with his opinion.

A year earlier, the president had argued the need for nuclear power in the *Daily Telegraph*. He felt that it was 'wishful thinking' for 'wind and solar to satisfy completely our seemingly insatiable appetite for energy …' As we saw in Chapter 7, his 'wishful thinking' is becoming reality in Germany. The 'wishful thinking' was Lord May's when he proposed that the UK should have new nuclear reactors by 2015 – now completely impossible.

I was disappointed also when I read the Royal Society's submission to the UK government's first energy review in 2006. I was very pleased to see the paper I had co-authored in *Nature Materials* referenced. Our paper had described how PV installations were rising exponentially in Germany. But I was disappointed to see that the report predicted PV would not contribute in the UK before 2015. As it has turned out the UK had nearly 2 GW of PV installed by the end of 2012.

On the other hand, the report offered the opinion that 'nuclear power stations can be built quickly and at reasonable cost'. The same paragraph explained that 'quickly' meant 15 years. They therefore suggested 2020 for the first new nuclear reactor to go on line, though it 'could be earlier'. At the present time 2023 is the date suggested for the first new 1.6 GW reactor. So, on sunny days, PV is already contributing more power to the UK grid around the time of maximum daytime demand than new nuclear may contribute in 2023.

I am sorry to say I have criticisms of other Royal Society reports. Their working group on 'Fuel cycle stewardship in a nuclear renaissance', published in 2011, makes a very important point: 'The UK's civil stockpile of separated plutonium undermines the UK's credibility in non-proliferation debates given this stockpile is the largest in the world and poses a serious security risk.' But I disagree with the reports recommendation: 'The UK's civil stockpile of separated plutonium should be reused as mixed oxide (MOX) fuel in a new generation of thermal light water reactors.'

The name 'mixed-oxide' fuel is a euphemism for the more accurate, but less publicly acceptable description: uranium fuel with added plutonium.

The report recommends a new plant be built at Sellafield to fabricate MOX fuel. It does not go into detail about the problems encountered by the first MOX fabrication plant, which was closed in 2011. The report simply refers to 'throughput problems' and 'failure to meet design expectations'. In fact, the original plant produced less than 2 per cent of its design throughput in its nine-year operating lifetime. The report does suggest how to ensure a better performance next time. It says the industry could learn from the French MOX experience. However, a MOX plant being built in the US by a French company is running well over design cost.

There is an alternative option for disposing of plutonium. It can be immobilised in a ceramic material, which is surrounded by radioactive waste to deter terrorists for a century and then buried deep underground. Evidence was submitted to the working party on this option by, among others, Allison Macfarlane who is now chair of the US Nuclear Regulatory Commission (NRC). The Royal Society report concludes that there are 'currently no technically proven and commercially deployable immobilisation technologies'. Yet the report's preferred option – a British MOX plant – can hardly be described as 'technically proven' or 'commercially deploy-able'. Contrast the report's opinion with that of the *Independent* in 2008 on learning of the MOX plant performance in the first five years of operation: '… one of the most comprehensive and catastrophic failures in British industrial history'.

The working party concluded that the MOX option 'provides an effective and technically proven management strategy for the stockpile'. But is MOX a *safer* management strategy than the status quo or immobilisation? The working party report does not compare the risks of these three options, though it refers to the status quo as a 'serious security risk'. I believe that moving nuclear fuel with added plutonium around the country and around the world is probably the riskiest of the three options, given the potential terrorist threat. I agree with the chair of the US NRC Allison Macfarlane and her colleagues who wrote in *Nature*, 'Britain should seriously evaluate the less costly and less risky method of direct plutonium disposal.'

THE POLITICS OF THE SOLAR REVOLUTION

The Royal Society report's main recommendation that a new generation of nuclear reactors be built to deal with the plutonium stockpile reverses the very sensible advice of another eminent source. Here are the wise words taken from the 6th Report of the Royal Commission on Environmental Pollution, chaired by a past rector of Imperial, the late Lord Flowers. They were written four decades ago, but are all the more relevant given what we have learnt since of the technical competence of some terrorists. 'Plutonium appears to offer unique potential for threat and blackmail against society because of its great radiotoxicity and its fissile properties', 'the construction of a crude nuclear weapon by an illicit group is credible' and 'there should be no commitment to a large programme of nuclear fission power until it has been demonstrated beyond reasonable doubt that a method exists to ensure the safe containment of long-lived, highly radioactive waste for the indefinite future.'

Recent UK governments have gone ahead with a new, large programme of nuclear fission power, while the location and the technology to be used for the long-term disposal of waste are still a matter of political and scientific dispute. Even if MOX fuel is used, the new reactors will still produce new plutonium for later generations to deal with.

My final comment on this working party report brings us back to my concern about the absence of a dedicated national PV laboratory in the UK. Recommendation 3 calls for the National Nuclear Laboratory to be 'fully commissioned'. I am happy if this recommendation means that more government money will be spent on solving the nuclear waste problem. But I am not aware the Royal Society has ever called for a national solar cell laboratory to be established, let alone 'fully commissioned'.

Given these experiences, it was with some trepidation that I approached the joint report of the Royal Society and the Royal Academy of Engineering in 2011, 'Shale gas extraction in the UK: a review of hydraulic fracturing'. This was a report about the risks associated with 'fracking' for methane, the villain we met earlier and the main component of natural gas. Fracking is a contentious issue in the UK and US, but the French government appears set against it.

Sadly, in this case too, I have concerns about the report. The two societies conclude that the extraction of shale gas 'can be managed effectively in the UK as long as operational best practices are implemented and enforced through regulation'.

The societies had been asked by the government to consider 'What are the major risks associated with hydraulic fracturing as a means to extract shale gas in the UK, including geological risks, such as seismicity, and environmental risks, such as groundwater contamination?'

In my opinion an even more important environmental risk than groundwater contamination is the impact of fracking on global warming. A number of relevant authorities support the view of the International Energy Agency that 'No more than one-third of proven reserves of fossil fuels can be consumed prior to 2050' if the world is to avoid a two degree temperature rise. This temperature rise would have disastrous consequences.

How big is the environmental risk from exploiting unproven reserves like shale gas? The joint societies' report admits 'climate risks have not been analysed.' It adds that 'the subsequent use of shale gas has not been addressed.'

The report does, however, discuss the important study of the carbon footprint of fracking by Robert Howarth and colleagues from the University of Cornell. These authors point out that there are significant leakages of methane to the atmosphere during the fracking process. The problem is that every molecule of methane that leaks into the atmosphere is a far more effective greenhouse gas than a molecule of carbon dioxide released when the methane is eventually burned. The royal societies' joint report states that 'The global warming potential of a molecule of methane is greater than that of carbon dioxide, but its lifetime in the atmosphere is shorter. On a 20-year timescale, the global warming potential of methane is 72 times greater than that of carbon dioxide.'

The report deals with the Cornell findings using the familiar 'more research' fudge: 'The same study recognised the large uncertainty in quantifying these methane leakages, highlighting that further research is needed.' The 'highlighting' in this conclusion was not a quote from the Cornell paper but it is the report's view. Howarth and his colleagues had plenty of data from the US on fracking. This enabled them to set quantitative limits on the methane leaks in practice.

The Cornell study estimated that between 3.6 and 7.9 per cent of the methane leaks into the atmosphere in the lifetime of the well. This was based on US experience. Even at the lower limit of 3.6 per cent, the Cornell team estimate that the greenhouse gas footprint of shale gas is higher than coal burning. This conclusion is not mentioned in the joint report.

Another environmental risk that the report did not refer to is that these methane leaks make the carbon footprint of electricity from shale gas even higher than natural gas electricity generation. Recall that in Chapter 8 we found that electricity generation from natural gas already had a carbon footprint nine times higher than the CCC limit.

The report downplays the risks of radioactivity being released in fracking. There are reports in the scientific literature of radioactivity in waste water from fracking in Pennsylvania. The societies' report tries to be reassuring. It points out that radioactivity 'is present in waste fluids from the conventional oil and gas industries'. I do not think this will reassure a local population threatened with fracking sites. They are unlikely to accept a conventional oil well, whatever the safety standards.

The societies were not asked to investigate alternatives to shale gas such as biogas from waste. The examples I have given of my concerns about the reports from the royal societies suggest we cannot expect much support for solar technologies. That is a shame, as there are some very strong forces opposing our revolution, as we will find in the next sections.

The nuclear lobby

Not all members of the nuclear lobby are against renewable power. I have enjoyed many public debates with nuclear supporters and some of my opponents have been in favour of the solar technologies. Their view is that new nuclear and the renewables should be 'part of the energy mix'. My 30 years' study of the UK energy scene convinces me that it is precisely because successive UK governments believe that nuclear must be 'part of the energy mix' that we lag so far behind Germany in exploiting PV and wind.

However, there are also many nuclear supporters who, quite correctly, see wind and PV as a threat to nuclear power. It is no surprise that workers in the nuclear industry are keen to see their jobs continue. They probably feel resentful at the way the environmental movement pillories their industry. My message for these members of the nuclear industry is that they should feel proud that their industry is tackling the *second* most important technological challenge for the future of our civilisation. You must have guessed what I believe to be the most important technological challenge. The second most important challenge is to find a safe and secure way to store the waste already produced by two generations of exploiting $E = mc^2$. This is a task that looks likely to take many generations to solve. It is particularly important challenge in this age of sophisticated terrorist activity. The industry should not be distracted from this task by building a new generation of reactors with new waste challenges.

Remember that in Chapter 4 we found that the UK government is already committed to underwrite the work necessary to find a solution for nuclear waste disposal to the tune of at least £3 billion a year indefinitely. The overall cost is estimated to be somewhere between £58.9 billion and £104 billion. These estimates do not appear to include the waste of currently operating second-generation reactors, which employ many of the current nuclear workforce. The government will have to underwrite this waste programme too. I conclude that, even if there are no more nuclear reactors, many nuclear workers are in the fortunate position of being able to count on a 'job for life'.

The nuclear industry clearly does have the ear of the UK and French governments at the present time. But how powerful a lobby are they? The ability of the nuclear industry to influence the UK government, and indeed the French government too, is weak because of its dependence on government support. In fact Areva, the only nuclear reactor manufacturing company left in Europe, and EDF, the company that wants to build Hinkley Point C, are both over 80 per cent owned by the French government. France has a socialist government, which appears more inclined than its predecessors to rein in the nuclear industry.

Most of the electrical supply companies that originally expressed an interest in building new nuclear reactors in the UK have withdrawn. This appears to be mainly because of the increase in costs due to the delays in construction of the two prototypes and the safety modifications required post Fukushima. Successive UK governments have insisted that new reactors would only be built if 'the market' wanted them. This has been shown to be a fiction as many green economists predicted. The coalition government's latest ploy – 'contracts for difference', which guarantee a wholesale price twice the current wholesale price during the 35-year lifetime of a new reactor – is a blatant subsidy.

We have seen in Chapter 7 that the peak wholesale price of electricity in Germany is already falling, and it will also do so in the UK as the contribution of PV and wind power rise. The difference between the high fixed contract price and the much lower wholesale price in 2023 will have to be paid by UK taxpayers.

THE HERE AND NOW OF THE SOLAR REVOLUTION

If the EU decides the 'contracts for difference' are not a subsidy and allows EDF to proceed, the crunch will come when one or other of the two delayed prototype reactors eventually works. The construction cost will rise yet again. There is always something to be learned from an engineering prototype, large or small, and these two prototypes are very big engineering projects indeed. Not only that, but the design has had to be modified because of the safety implications of the Fukushima disaster. The first EDF reactor will therefore itself be a prototype. At that stage, the words 'straw' and 'camel's back' will become appropriate. Any industry that depends on a blank cheque from one or more governments is not in a strong position.

The fossil fuel lobby

When the solar revolution finally takes hold in all countries and the definitive history is written, I think the researchers may well conclude that the strongest lobby in the counter-revolution was the one we have yet to discuss: the fossil fuel companies. Their influence will probably turn out to be more important than governmental support for nuclear or the indifference of the military.

It has been my experience that any enthusiasm shown by oil companies for renewable energy has waned as the technologies have taken off and become a potential threat to their core interest, which is selling oil and gas. Take BP Solar for example, a small division of the leading multinational oil company responsible for the Deepwater Horizon disaster in April 2010.

I have before me a copy of a BP press release from 1998 in which the then BP Group Chief Executive, John Browne, announced ambitious targets to grow their subsidiary, BP Solar, to a $1 billion turnover by 2008.

At the time, BP Solar had an active research laboratory south-west of London and my Quantum Photovoltaic group at Imperial had a fruitful relationship with them. The head of silicon research and development at BP Solar, Tim Bruton, gave lectures to our masters students for some years. The company was doing well, and Tim reported that in 2000 BP Solar was a world leader in silicon panel sales. Sadly this position was not maintained. By 2003, not long after we had successfully completed a research project with them, the BP Solar laboratory was moved to Madrid. BP's European solar cell production line had moved there a few years earlier. Many researchers left the company at that stage.

In 2002, while I was in the US, Tim arranged for me to visit the new BP state-of-the-art solar panel plant in California. It was one of the first manufacturing plants to produce second-generation, thin film cells made of the semiconductor cadmium telluride (CdTe). The day before the visit, I heard that it was cancelled. BP Solar had decided to close the plant. In 2009, by which time BP Solar had dropped well down the solar manufacturing league table, a US company, First Solar, became the first to manufacturer 1 GW of solar modules in a year and also drop below $1/W production cost. So First Solar achieved the level that John Browne set for BP solar, one year after his target date. Ironically, they did so with CdTe cells, the same material that BP Solar were pioneering, but abandoned, seven years earlier.

It was no surprise when in 2011 BP Solar finally decided to stop making PV panels. Clearly, they were not prepared to commit to the scale of production at which their cells might compete in a mass market. Solar panels had always been a side issue for BP. To put this in context, in 2011 First Solar announced an impressive $2.8 billion of net sales. In the same year, BP group sales of fossil fuels were $376 billion, which is 134 times higher than First Solar. In 2011, when BP Solar took their decision to end PV panel manufacture, their expenditure on oil and gas development was $20 billion.

I expect most of the other fossil fuel companies have a similar ambivalence to the growing renewable industries. The most imminent threat fossil fuel companies face to their profits is electricity from wind power, PV, biogas and the other renewables as they replace electricity from coal and gas burning.

I mentioned in Chapter 8 that the coalition government in the UK is prevaricating over setting a strict carbon limit for electricity generation in 2030. I assume this is as a result of lobbying by the fossil fuel industry claiming that natural gas is cheaper than the renewables and shale gas has a lower carbon footprint than coal. We need to get over to our governments that both these claims are myths.

The fossil fuel lobbies are powerful in most countries. According to the Renewable Energy Association, the fossil fuel industry in the US spends over 20 times more on lobbying politicians than the renewable industry. This could be an underestimate. We saw that BP, a typical large multinational fossil fuel company, has turnover 134 times higher than one of the largest PV manufacturers.

In Britain the fossil fuel industry is omnipresent. In 2013 the *Independent* announced that the Prime Minister, David Cameron, had appointed a former lobbyist for British Gas to be his personal adviser on energy and climate change. Later that year the government admitted that eight employees of fossil fuel companies had been seconded to DECC and were helping to draft energy policy. Whilst no doubt these individuals carry out their tasks independently, their background gives the impression of bias which the government should try to avoid.

How can we hope to influence government and the fossil fuel industries to implement green policies? I have some suggestions which I will present in Part III when I set out a manifesto for the solar revolution. However, it is important to point out now that I believe fossil fuel companies, in all countries, have an Achilles' heel: fossil fuel subsidies.

According to the International Energy Agency, subsidies to the fossil fuel industry worldwide are six times higher than subsidies to renewable industries. In the UK in 2010, subsidies to the UK fossil fuel industry were £3.63 billion whereas subsidies to the renewable industry were only £1.4 billion according to the Renewable Energy Association. By 2013, a study commissioned by the UK House of Commons Environmental Audit Committee concluded that the subsidies for natural gas alone were £3.6 billion a year whereas for all the renewables taken together were only £3.1 billion.

I believe that these latest subsidy figures provide very useful ammunition for the environmental movement in the UK. If the gas subsidies were transferred to renewable biogas generation they would make a major difference to the latter. Clearly the current subsidy to biogas produced from anaerobic digestion and other wastes is only a very small part of the smaller renewables pot. The stimulation of the new subsidy will reduce the cost of biogas supplied to the gas grid and for generating electricity. This answers the argument of the natural gas lobby that if they lose the subsidy, consumer prices will rise.

The fossil fuel lobby may also argue that they produce much more gas than the renewables currently do, so they need a bigger subsidy. The answer is that they are clearly an established industry and should not need subsidies. We should be subsidising our indigenous sources of biogas from waste rather than fuel from foreign fossil fuel companies. Not only are these indigenous sources more secure, they have a far smaller carbon footprint.

Of course you will be aware that there is another major component of government fossil fuel subsidy: support for the exploration and exploitation of oil for transport. The International Energy Agency has calculated that 'electricity and heat production' is the sector that contributes most (41 per cent) to global carbon emissions. The second most important sector is transport (22 per cent). It is high time we moved on to consider how the solar revolution can lead us to the holy grail of carbon-free transport.

III

The Future of the Solar Revolution

ELEVEN

Foretelling the Future

Prediction is very difficult, especially about the future.

Neils Bohr

We already have the technology to ensure that all our electricity is generated from the sun. An all-solar electricity supply is a necessary condition for avoiding the worst of global warming; but it is not sufficient. The fossil fuels that power our cars, trucks and aeroplanes make the next biggest contribution to global warming after electricity and heat generation. We urgently need to start making major reductions in the amount of carbon dioxide emitted by transport. So, before we look at my manifesto for the solar revolution, and consider how the revolution might develop round the globe, I want to introduce you to the existing technology, and the research and development of new technologies that can decarbonise our transport.

One low carbon technology, the electric car, is already performing well. Some future developments, such as fuel cells, will encourage more people to use electric cars. Fuel cells will extend the range of the electric car and reduce the time spent refuelling them. We will also meet solar fuels and the challenge of producing them from carbon dioxide in the air and sunlight on our own rooftops.

Will the solar revolution develop fast enough to avoid the catastrophic effects of global warming, oil depletion or nuclear disaster? As Neils Bohr says, predicting the future is not easy. Scientists regard experimental results as the arbiter of contentious issues and the best test of a prediction. So let's begin by testing the accuracy of predictions made in 1979 about the future of our energy supply. These particular predictions were important for me personally. They were influential in my change of research career.

2079: a century of technical and sociopolitical evolution

My conversion to the solar cause came as a dramatic revelation. In 1979, I experienced one of the lowest points of my particle physics career while at CERN, in Geneva. I could recall many exhilarating times there, but the problems of working at a laboratory 800 kilometres from my family and university had finally taken their toll. I had found being spokesperson for a five-nation, 30-physicist collaboration stressful. The next experiments at CERN would involve 500-physicist collaborations and even more politics and commuting.

I wandered the CERN corridors aimlessly, until I found myself outside the library. In the past, I had spent many happy hours there. I decided I could do worse than look at some papers. Perhaps I might find some guidance on a new research field? Browsing the racks of preprints and the latest journals, I came across a collection of papers from a conference held that year, 1979, to commemorate the works of Albert Einstein. Had he lived, the great man would have been 100. That meant Einstein was only 26 when he started both quantum theory and relativity in the same year. I was then 36. I realised I had better get started if I was serious about a new research career.

Standing at the racks, I skimmed through the list of contents marvelling yet again at the unique breadth of Einstein's contributions to physics. My eye was caught by the observation that the last paper in the proceedings was of a completely different nature to the others. It was titled '2079: a century of technical and socio-political evolution' by David Mathisen, which you can find in the journal *Impact of Science on Society*. I started reading, was captivated, slumped to the library floor and drank in every word.

In summary, the paper described the history of the world from 1979 to 2079 as dominated by problems associated with the unequal distribution of energy resources and the environmental impact of fossil fuels and nuclear energy. Eventually, in 2028 according to Mathisen, 'with much of the world teetering on the brink of anarchy', the solution would come with a breakthrough called the solar/laser/hydrogen system. This revolutionary technology enabled solar electricity to be exploited to generate fuel for transport, heating and industry.

I was transfixed; solar electricity was the answer. I decided there and then that I would switch my research career to solar photovoltaics, though it took a decade to complete the change.

Rereading the article in 2014 I am struck by how many predictions Mathisen got right. He foresaw global warming as a problem. He expected coal to be the main pollutant, which it still is in many parts of the world. He did not foresee the major expansion in natural gas. This has a lower carbon footprint than coal but nowadays makes a major contribution to global warming.

Mathisen even predicted that there would be a disastrous terrorist explosion 'In the early years of the (21st) century' in New York. He did not predict airliners but rather a makeshift nuclear weapon made from 'pirated shipments of plutonium'. My reaction on rereading this prediction is of relief that, as yet, we have been spared a terrorist nuclear atrocity. Fortunately, the plutonium economy has not developed as fast as Mathisen feared. However, the international and intra-national transport of plutonium-containing fuel does happen. The danger will intensify if new nuclear reactors use MOX fuel. What is particularly worrying now, as Mathisen correctly predicted, is the increased sophistication of terrorist groups. Remember Jungk's warning about nuclear fission in 1958, which we met in Part I. He said that most scientists of the 1930s were so preoccupied with the enormous strength of the force holding the nucleus together they could not foresee that just one low energy neutron could penetrate a uranium nucleus and split it apart. Preoccupation with nuclear weapons has blinded our political, military and scientific leaders to their uselessness in the face of just one single suicide bomber who can penetrate our defences and cause devastation.

It was not just the timing and sophistication of the New York terrorist disaster that Mathisen got impressively right. He also anticipated that, by involving plutonium, it would have some relation to the world's energy preoccupations. It is possible that 9/11 had some connection with oil issues given that most of the terrorists originated from Saudi Arabia.

Mathisen also predicted two related terrorist outrages around the same time, though he got the locations wrong. He said Brussels and Moscow rather than London and Madrid. It is possible these two atrocities also had energy links. The London and Madrid bombs were probably retaliation for involvement in the Iraq invasion of 2003. There are those who believe the invasion had a connection to oil depletion.

Finally, Mathisen predicted a reactor explosion in 2008 that finished the Nuclear Age. Only three years out in 32; that is impressive. He set the accident in France, rather than another state with a large civil nuclear programme, namely Japan. Mathisen accurately predicted reactor meltdown due to an explosion resulting from 'the pressure of the steam produced by the supposedly foolproof emergency water-cooling systems'. We have yet to see if the Fukushima disaster will finish the Nuclear Age, but I think it is a sizeable nail in its coffin.

I like to think that there is still time for Mathisen's solar/laser/hydrogen breakthrough in 2028 to be a forecast of the solar/artificial leaf/methanol system that we will meet in Chapter 13. However, I don't think the world can wait until 2028 to introduce such a system. Fortunately, it probably can be developed before then. We already have solar cells that can convert sunlight to electricity with 42.5 per cent efficiency. I believe they could be an important part of the system. Mathisen predicted the 2028 breakthrough would came from a US–Italian research collaboration. The 42.5 per cent efficient cells were developed by a company founded by two British physicists and an Italian that was taken over by a US company.

The crucial point that Mathisen was making way back in 1979 was that our environmental problems will not be solved until we find a way to generate a replacement for the fossil fuels we use for transport and for industrial processes. I will share with you some ideas my colleagues at Imperial are working on to do just that, in Chapter 13.

Later I make my own predictions for the way the revolution will develop up to 2028. I hope my predictions over a decade and a half will be as accurate as Mathisen's have proved over the past 34 years.

First, I will describe a low carbon transport technology that is already delivering: the electric car.

TWELVE

The Electric Car

What Englishman will give his mind to politics as long as he can afford to keep a motor car?

George Bernard Shaw, *The Apple Cart*

In 1979 David Mathisen predicted that by 2028 much of the world will be 'teetering on the brink of anarchy' without a solar-generated replacement for fossil fuels. This chapter will discuss two options for low carbon transportation, the electric car and the fuel cell. Electric cars are already on our roads and some buses are powered by fuel cells in our cities. Electric cars with fuel cells are being developed. They should increase the popularity of electric cars as they will increase their range and reduce the time taken to refuel.

The price of an electric car, the amount of government subsidy and the availability of recharging points for the battery are major factors that will influence the speed at which the market will take off. If the range of an electric car is extended by a fuel cell, the availability of refuelling stations for the cell will also be a factor. We will look at the possibility of recharging the battery in the electric car from a PV system on your roof at home. In the next chapter, we will look at a possible future development: solar fuels made from carbon dioxide extracted from the air. Fuel cells are already available that could convert these solar fuels to electricity to drive your electric car.

But first let's look at the electric car and what it can achieve on its own, without a fuel cell to extend its range.

THE ELECTRIC CAR

The history of the electric car

The early 1900s marked the start of the quantum revolution and the development of wireless communication. The motor car was also developing fast at that time. In 1900, electric-powered cars outsold all other types in America. Electric car development had started 70 years earlier, powered by electric motors using Faraday's ideas.

There were three main problems for electric cars in the early 1900s: expense, restricted range and Henry Ford, who mass produced cars powered by petrol engines. Today, a range of electric cars are available, but they are expensive, still do not have the range of an internal combustion engine and they face stiff competition from the commercial descendants of Henry Ford.

The battery is the crucial component of the electric car and the major reason for its high cost. Most electric cars nowadays have lithium batteries, which are larger versions of the batteries in laptop computers and mobile phones. Research and development programs are under way to produce batteries that are cheaper, smaller, lighter, store more electrical energy, are quicker to recharge and have longer lifetimes. The challenge is that these performance requirements imply very different physical and chemical properties for the materials. Satisfying them all in one cheap and safe car battery is not easy.

Given these problems, car manufacturers are building hybrid cars with both electric and petrol motors, rather than pure electric cars. Their aim is to develop generations of hybrid cars in which the power provided by the electric motor gradually increases and the power of the petrol engine decreases. This seems to me to be a policy designed to maintain our reliance on petrol for as long as possible. It suggests that the automobile industry takes a similar view of the all-electric car as Saint Augustine's famous prayer about chastity: 'Grant me an all-electric car, Lord, but not yet.'

I believe we could reduce our carbon footprint faster if higher priority was given to the all-electric car than the petrol/electric hybrid. Adding a fuel cell to the electric car produces on-board electricity for the electric motors, extends the range of the car and reduces the size of the battery. Amory Lovins, the American energy efficiency expert, was the first to point out that a lighter, more aerodynamic body makes hybrids and electric cars cheaper by making their costly batteries or fuel cells smaller. This speeds their adoption and enables a virtuous spiral of still lighter weight and lower cost. A light electric car can still have the strength you need for safety, as Lovins has demonstrated.

Another benefit of the electric car is that while in the garage it provides electricity storage for the household PV system. I will take this argument further in a later section and explain how a third-generation PV system on the roof of a house can charge the electric car in the garage and also provide much of an average family's household electricity. For this application, it would clearly help if batteries were cheaper and lighter, so they could be easily exchanged. A second battery, or part of the battery, could then be on charge at home while the car is in use.

But let's first look at ways in which present-day electric cars are becoming more practical and popular with some drivers.

Today's electric cars in practice

I had an extremely encouraging discussion with solar revolutionary Kevin Sharpe, the founder of Zero Carbon World. This is a charity based in Bath not far from where I live.

Zero Carbon World freely donates charging stations for electric cars to the hotel and leisure industry. They supply restaurants with systems that can recharge in 45 minutes and hotels with systems that recharge more slowly to higher levels. The idea is that the service will be free if the driver eats in the restaurant or stays a night at the hotel. If only a recharge is required, the price will be £5. Depending on the length of time, the recharging could be cheaper than parking in some tourist areas. The Best Western hotel chain had installed 82 charging points by the end of September 2013. Best Western aim to go global with the service. Yes, I did ask. About a quarter of Zero Carbon World's stations get their electricity from an all-renewable electricity provider.

As the number of such stations rises, long-distance electric car journeys for vacations and journeys that can be planned ahead of time are going to become increasingly practical, pleasant and certainly cheap. Kevin describes a recent trip, Bath–Vienna–Geneva and back, that covered 4,000 miles in all. His electric car typically goes four miles for each kWh of electrical energy consumed. Assuming an average electricity cost of 10p for each kWh, one might expect to spend £100 on electricity during the trip. Had he been driving a petrol car the trip would have cost over £590.

In fact, the cost of Kevin's 4,000-mile round trip was *zero*. He started with a charge from his rooftop PV system at home and the recharging stations in France, Germany, Austria and Switzerland were all free – like the ones Zero Carbon World is installing in the UK.

Kevin's European trip is an example of not just how cheap but also how fuel efficient electric cars are. One reason for their efficiency is that when an electric car slows down it generates electricity, which can be stored in the battery. Remember that we found in Chapter 2 that an electric motor can act as a dynamo. As use of the electric car expands, the road safety advice for drivers entering built-up areas could be extended: cut your speed to save lives … and to generate electricity.

Recharging your electric car at home

Sales of the electric car should get a major boost when it becomes practicable to recharge an electric car by rooftop PV. The prospect of free fuel for 25 years would be a huge boost for both the electric car and PV technology.

It can be done already with current PV technology, even in the rainy UK. But you need a second electric car or a spare demountable battery to be left on charge when the sun is shining. Sadly, today's electric car batteries are expensive and difficult to demount. Kevin Sharpe has demonstrated the practicality of rooftop PV recharging for a two-car household. He points out that it is also practical for a one-car household where the breadwinners work at home. This is an increasingly common phenomenon.

Take Kevin's figure of four miles' driving for each kWh of electrical energy and assume a typical 2.5 kW power, domestic rooftop PV system in the UK. On average, the sun shines for 844 hours a year in the UK. Multiply the 2.5 kW power and the 844 hours together to get the total 2,110 kWh electrical energy generated by such a PV system on the average rooftop in the UK. Then multiply again by Kevin's four miles for each kWh and you find a yearly total of 8,440 miles provided by the rooftop system per year. This is very close to the 8,430 miles per year which the UK Department of Transport National Travel Survey estimates was the average annual domestic mileage in the UK in 2010. In countries nearer to the equator than the UK, a system of the same power would produce a greater mileage.

In practice, Kevin reports that he gets a significantly higher mileage from his system than my average estimate, even when we have allowed for the difference in the power of the two systems. The Bath region is quite a bit sunnier than the UK average. Also this enhanced performance could be due partly to having PV cells on both the south-east and the north-west elevations of his roof.

Hence the sun falls on at least part of his system more often when the car is on charge.

Taking all the annual mileage for free from the domestic rooftop will mean nothing is left for the household. But, remember the third-generation CPV cells with three times the efficiency of the cells on rooftops today? If the 42.5 per cent CPV cells were deployed on rooftops, they could, in principle, generate three times the electrical energy of present-day PV panels *covering the same roof area*. To make the price affordable, they would have to be deployed in a sunlight concentrating system, which would reduce the overall efficiency. We will later meet small CPV units which could track the sun inside a panel looking like present-day technology. Also, in the UK, much of the sunlight is diffuse, rather than coming directly from the sun. It is only the direct sunlight that can be focused by a high concentration system. My old group at Imperial, and others, are working on ways to concentrate the diffuse as well as the direct sunlight.

For all these reasons, an affordable CPV system based on 42.5 per cent technology will probably only have around twice the efficiency of today's panels. That means this third-generation system covering the same area of roof as present-day technology should be able to generate double the electrical energy. Therefore this system would deliver the same amount of electrical energy for household use as today's technology plus enough electrical energy for the average yearly domestic car mileage.

I really don't think it would take long for domestic roof CPV systems to be developed and the 42.5 per cent cells are already in production. I think a cheap and easily removable spare battery may take rather longer. So this rooftop recharging station will mainly be useful for households with breadwinners working at home.

Fuel cells

The distance electric cars can be driven between recharging will increase and the time for recharging will reduce as battery technology improves. Also, the number of recharging points is increasing thanks to solar revolutionaries such as Zero Carbon World. But the limited range and the time taken to recharge are fundamental problems for many car owners. It is even more of a problem for heavier vehicles, such as lorries and buses, which require more powerful electric motors and even bigger batteries. Another way to power the electric car, which is already in use in some heavy vehicles, is the *fuel cell*.

Like other technologies of the solar revolution, the fuel cell was invented by a Victorian, but not commercialised until it became a strategic necessity in the US. The main difference from our other inventors is that William Grove was Welsh. Like some of his scientific contemporaries, he did have independent means; he was a barrister and became a judge.

The early development of the fuel cell was funded by the US space programme. Fuel cells provided on-board electricity for the Apollo missions to the moon and later for the space shuttle.

Unlike the internal combustion engine, a fuel cell is not a burning technology. It is much less complicated and significantly smaller and lighter; this was the great advantage for NASA. A fuel cell runs on a fuel made up of much simpler molecules than petrol, such as hydrogen, methanol or ethanol. A well-controlled chemical reaction with oxygen from the air converts the fuel into electricity and waste products. If the fuel is pure, there are far fewer types of waste than in the exhaust of an internal combustion engine. In the case of hydrogen, the waste product is water; for methanol and ethanol, the waste consists of water and carbon dioxide. We will see that fuel cells can have very low carbon footprints, but I think it also a very great benefit that they will vastly improve the quality of the air in our cities.

THE ELECTRIC CAR

A fuel cell is nearly as simple as a solar cell and similarly has no moving parts. It is built with a plastic membrane or a ceramic material sandwiched between two electron-conducting contacts known as *electrodes*. A wire from one electrode passes through an electric motor to the other electrode.

The importance of the plastic membrane and the ceramic material is that they only allow one type of charged particle, usually positively charged protons, to pass through. Remember that the nucleus of the hydrogen atom consists of a single proton. Removing the electron from a hydrogen atom liberates a proton.

In a hydrogen-powered fuel cell the hydrogen passes over one electrode, which has a special coating. The coating is an example of a *catalyst*. This is a substance that speeds up a chemical reaction without being changed itself. In this case, the reaction strips electrons from the hydrogen atoms. The liberated protons are free to pass through the membrane or ceramic. As a result, large numbers of electrons build up on this electrode. They all have the same negative charge, which, I hope you remember, means they repel each other. The electrons try to get as far away from each other as possible. However, there is nowhere for them to go, apart from along the wire towards the electric motor, forming an electric current. The current flowing in the magnetic field of the motor turns the coil and so produces useful electric power.

The electrons then continue down the wire to the second electrode. This electrode has another special coating, which is a catalyst that promotes a second chemical reaction. Electrons from the wire and protons that have passed through the membrane react with oxygen from the air to produce water. This is the only waste product if the hydrogen is pure.

Hydrogen fuel cells in electric cars

Fuel cells are a greener and lighter option for extending the range of a hybrid electric car than small petrol engines. A fuel cell hybrid is also a more elegant option. The battery can be smaller and the car lighter. Either battery or fuel cell can feed electricity to the same electric motor under automatic control. Fuel cell powered buses are now running on the streets of many cities. In 2013 Hyundai claimed that their ix35 Fuel Cell would be the world's first mass-produced hydrogen fuel cell vehicle. They say that they will be producing 1,000 in 2015.

There are still a number of problems with developing a 'hydrogen economy', which have led to controversy among analysts about the economics of the car and the belief that the growth in the take-up of hydrogen-electric hybrid cars may be slow. As with all new technologies, the initial costs will be high. The development of cheaper catalysts would help. Then there is the need for a network of hydrogen refuelling stations. These will be more expensive than battery recharging stations, in particular because of safety issues.

A start has been made and there is progress towards hydrogen refuelling station networks in Europe, Japan and Korea. Refuelling takes around five minutes, which is much quicker than battery recharge, but still longer than a petrol station.

One particularly important question for the solar revolution is how green is the source of the hydrogen. The largest reduction in greenhouse gas emissions would be achieved if the hydrogen were generated from renewable energy. The technology exists to do this. It is known as water splitting by electricity or the *electrolysis* of water to give hydrogen and oxygen.

Electrolysis is yet another important application of electricity to which Michael Faraday's experimental researches contributed. It is essentially a hydrogen fuel cell operating backwards. Remember, in the hydrogen fuel cell a chemical reaction at one electrode produces protons which pass through a membrane and electrons which form a current in the external circuit. At the other electrode the protons from the membrane and the electrons from the wire react with oxygen from the air to form water.

In the case of electrolysis, the two electron-conducting electrodes are immersed in a water solution that conducts electricity. There is a plastic membrane between them. The membrane ensures that protons only go one way, just like the one in the fuel cell. In this case, the plastic membrane has another very important function: it keeps the hydrogen and oxygen gases apart as the water decomposes. Another difference from the fuel cell is that the wires connecting the electrodes pass through a generator of renewable electric power. Both electrodes have catalytic coatings that speed up reactions that are the reverse of those in the hydrogen fuel cell.

The generator pulls electrons from the water molecules in contact with one catalyst and pushes them down the wire to the other electrode leaving protons behind in the water. The protons pass through the membrane while oxygen gas bubbles off. At the other electrode, the catalyst enhances the reaction between the electrons coming down the wire and the protons coming through the membrane to form hydrogen gas that bubbles off this electrode.

The cost of the hydrogen fuel then depends on the price of the renewable electrical power, the cost of the metal electrodes and catalysts and the energy expended in compressing and storing the hydrogen gas safely. In the next chapter, we will find how the reactions can be speeded up, and the economics of solar fuel can be improved, if the metal electrodes are replaced by semiconductors and illuminated by our super-hero, the photon. This is done in what is known as a photo-electro-chemical cell (PEC).

Another major challenge for the hydrogen-electric hybrid car is one of public perception of the risk. The role that hydrogen played in the tragic history of the airships is ingrained in the public consciousness. Of course, good engineering can ensure a very high degree of safety but I share the concerns of many non-scientists who are sceptical of statements by technologists when safety and commercial interests become entwined. Just think thalidomide, GM foods and nuclear power. I am happy riding on a bus powered by a hydrogen fuel cell. The bus is heavy, solidly built and unlikely to be involved in a high-speed accident. Call me over-cautious, but I doubt I would be the only person concerned about sitting in the back seat of a car on top of a pressurised hydrogen fuel tank.

In the next chapter, we will meet research which may eventually lead to the production of solar fuel from PV on a domestic rooftop. I certainly would not be happy about running a hydrogen refuelling station in my garage.

What we need is a safer green fuel for the range extender in the family electric car. Time to introduce you to solar fuels that can be produced from carbon dioxide in the air.

THIRTEEN

Solar Fuel

Life exists in the universe only because the carbon atom possess-
es certain exceptional properties
James Jeans, *The Mysterious Universe*

The popularity of electric cars would increase if their range could
be extended, and refuelling time reduced, by adding a fuel cell
that was inherently safe. Demand would expand even further if we
could produce the fuel itself on our rooftops at home from carbon
dioxide in the air and sunlight. The promise of free solar fuel for
life might convert the most avid devotees of the internal combustion engine, even the many followers of the BBC programme *Top
Gear*, to the joys of an electric car.

There are a number of technologies that could eventually
produce solar fuel from PV on the domestic rooftop and carbon
dioxide from the air. We will look at two of them.

In the last chapter, I expressed concern that safety issues might
rule out hydrogen as a solar fuel for domestic generation. Instead
of hydrogen, I propose two other candidates: methanol (methyl
alcohol) and ethanol (ethyl alcohol). Both have a much simpler
chemical structure than petrol. Fuel cells that convert them to
electricity are available commercially. Both are already produced
from biomass. Hence hybrid cars can be introduced here and now
without waiting for the solar fuels to be developed.

I was pleased to learn that in July 2012 two companies,
ECOmove and Insero E-mobility started collaborating to develop

a hybrid electric car with a methanol fuel cell. They are using methanol fuel cells produced by the company Serenergy, which manufactures fuel cells that have very impressive efficiencies up to 57 per cent. Appropriately this is being done in Denmark, which pioneered two other green technologies: wind power and local combined heat and power schemes. The companies claim the hybrid will run for 800 km between fill-ups. If the project is successful, battery sizes will be smaller and cheaper and the car will be much lighter than a petrol hybrid.

Both ethanol and methanol are already used in internal combustion engines, in their pure forms and also mixed with petrol. Ethanol fuelled cars are very popular in Brazil. Methanol and ethanol are used to power racing cars in the US.

As they are both liquids, an electric car with a methanol or ethanol fuel cell could be refuelled more quickly than with hydrogen. The fuels are also easier to transport than hydrogen, do not leak from containers like hydrogen gas, do not require expensive pressurisation and are safer.

Methanol and ethanol do have safety issues, however. We are all aware of the adverse effects of excessive ethanol consumption. Though methanol occurs naturally in the human body, and in our diet, drinking a straight 25 ml or more may be fatal if not treated within 10 – 30 hours. This is one of the few cases where methanol is less safe than petrol where the comparable number is 120 ml. One well-known use of methanol is to discourage human consumption of cheap and readily available ethanol by denaturing it as 'methylated spirits'.

In general, however, both methanol and ethanol are safer than petrol. The concentrations of methanol and ethanol vapour in air have to be much higher than petrol for the vapour to ignite. Flames burn more slowly and with less heat than petrol. Never fill a methylated spirit burner with petrol!

Unlike petrol, both ethanol and methanol readily mix with water. So an ethanol or methanol fire can be dowsed with water.

This will also mean that after the fire, or after spillage, the ground pollution will be less as the alcohol will be diluted.

The 'Ethanol Economy', in which ethanol replaces the fossil fuels on which so many of our industries are dependent, is being promoted by the Nobel Prize-winner George Olah, together with Alain Goeppert and Surya Prakash. The three have written a book, *Beyond Oil and Gas: The Methanol Economy*, arguing that this would be more versatile, safer and more economic than a 'Hydrogen Economy'.

As with hydrogen, the first issue to be addressed is whether methanol and ethanol can be produced in such a way that the overall carbon footprint is very much lower than petrol. This can be achieved if two conditions hold: first, all the carbon atoms in the alcohols must originate from carbon dioxide extracted from the air; second, no new carbon dioxide molecules can be emitted into the air during fuel production. This would be the case if the fuel preparation only involved renewable power. If both these conditions hold, use of the fuel to generate electricity in a fuel cell will be a *carbon neutral* process. The carbon dioxide, which is a waste product in the alcohol fuel cell, would contain the same number of carbon atoms as were extracted from the air originally. A fuel that in production follows these two conditions would be an ideal *solar fuel*.

Solar fuels should be distinguished from biofuels produced from biomass. In the latter case, though the carbon atoms have been extracted from the air in the months before harvesting, some carbon dioxide is emitted in the preparation of the fuel. If this can be minimised, then methanol and ethanol from biomass could be used in fuel cells of electric hybrid cars while solar fuels are being developed.

In addition to being carbon neutral, solar fuels will have the additional advantage over biofuels that they will be produced with high purity. This is important for fuel cells as impurities degrade the catalysts. In addition, some impurities react with the oxygen and end up being emitted as greenhouse gases.

Can solar fuels be produced from carbon dioxide in the air?

Though the increasing amount of carbon dioxide in the atmosphere is extremely worrying, the absolute amount is quite small, around 400 molecules in every million molecules of air. The pre-industrial era level was 280 molecules in every million, so humankind has made a major relative change in the amount of CO_2.

Extracting such a small proportion of carbon dioxide from air is not easy. In fact, a report in June 2011 from the American Physical Society (APS) called 'Direct Air Capture of CO_2 with Chemicals' concludes that direct air capture (DAC) 'is not currently an economically viable approach to mitigating climate change.' Dare I disagree with yet another learned scientific society? Here are three points that support my case.

One of the reasons the semiconductor industry has been so successful is because it follows an important approach to novel devices: first, demonstrate the device performance, then decide how to mass-produce it in an 'economically viable' manner. The APS report does not appear to be following that approach. It points out that 'No demonstration or pilot-scale DAC system has yet been deployed anywhere on earth'. I believe it is therefore too early to decide on economic viability.

Second, in Chapter 6 we met another reason why the semiconductor industry has been so successful. In many cases, the military or NASA paid research and development costs of new ideas; then the industry made the devices 'economically viable'. Fortunately, in the case of carbon dioxide capture from air, the military and NASA have already played a part. Service personnel and astronauts may well be super-persons, but they breathe out as much carbon dioxide as you and I. Life in a nuclear submarine or a space station would be intolerable had the technology not been developed to remove carbon dioxide from the recycled air and keep the concentration near the level we breathe on the ground.

The third reason I believe research will eventually find a viable way to extract carbon dioxide from the air is because there is a sunlight-powered system that already does the job effectively and produces carbon-based fuels. Every time I mow the lawn and observe the weeds thriving faster than the grass, I am reminded that nature found a way to use sunlight to capture carbon dioxide from the air and turn it into the chemical fuel that weeds need to grow: photosynthesis. I agree it is slow, but it is fast enough to spoil my lawn and all too 'viable' in most senses of the word.

As I described briefly in Chapter 1, photosynthesis produced all our fossil fuels. The oil industry describes these as 'economically viable', though burning them all will be incompatible with the survival of our civilisation. We should think a bit more about how nature uses sunlight to turn carbon dioxide from the air into fuel.

Can solar fuels be produced as a by-product of solar electricity?

The APS report does not consider the use of renewable technologies or the generation of solar fuels: '… the report does not investigate roles for synthetic fuels derived from CO2 captured from air …' This was outside the scope of the report, though it could impact on economic viability. They also conclude that the use of renewable power sources to capture the carbon dioxide 'would usually be less cost-effective than simply using the low-carbon energy' to replace high carbon power generators.

My view is that photovoltaic electricity can perform both tasks – replace high carbon power generation *and* generate solar fuels. The latter can be a by-product of the former.

Let me explain this by describing the ideas of Aldo Steinfeld, whom I met at a summer school in Sicily in 2012. Aldo and his team at the Swiss Institute of Energy Technology at ETH Zurich have developed a nanostructured gel that will absorb carbon dioxide at ambient temperatures and release it again, with high purity, when the gel is warmed to a temperature of 90 °C.

I think these results are very exciting. The QuantaSol version of the triple junction cell can be designed to operate with 40 per cent efficiency above 90 °C in a concentrating (CPV) system. The 40 per cent efficiency means that 40 per cent of the power in sunlight is converted into electrical power. What happens to the 60 per cent of the power in sunlight that isn't converted into electricity? It is wasted in many CPV systems. Most of it ends up as thermal energy: the jiggling of the atoms in the solar cell. That is something the manufacturers of CPV systems try to avoid because, as solar cells get hotter, their efficiency falls.

One of the clever ways concentrator manufacturers keep the cells cool is to pump water through pipes immediately behind the cells. The cooling of the cells warms the water. If our cells are operated at 110 °C (we have tested them at temperatures nearly as high and they work well) then the water should reach 90 °C and can be used to release the carbon dioxide from Aldo's nano-structured gel. The water cooling a concentrator system typically ends up less than 20 °C lower than the temperature at which the cells operate.

Using the waste heat of a CPV cell is a further example of a combined heat and power system (CHP) which we met earlier in Chapter 8. How much power can we get from the waste heat in the cells? Any physics undergraduate will be able to show you how to calculate the efficiency with which useful power can be extracted using the second law of thermodynamics that we met in Chapter 9. Recall that the efficiency depends on the hot and cold temperatures. If a CPV cell is at 110 °C and the ambient temperature is 25 °C then, according to Sadi Carnot and Lord Kelvin, useful power can be extracted with up to 22 per cent efficiency.

SOLAR FUEL

So from the 60 per cent of the power in sunlight wasted as heat we can hope to extract 22 per cent as useful power. That means we can extract 13 per cent (0.6 times 0.22) of the total power in sunlight as useful power from the waste heat of CPV cells.

Does the 13 per cent efficiency ring any bells? In Chapter 10 we found it was an efficiency that was typical of the solar panels on rooftops today. So let's summarise what a rooftop CPV system with QuantaSol cells operating at 40 per cent efficiency at 110 °C could achieve. In most locations it could produce around twice the electrical power of today's technology from the same area of roof. In addition there will be useful power from the waste heat which would be equal to the electrical power of today's rooftop technology. In all, the CPV system could produce about three times the power of current PV panels.

What could we do with all this power on our rooftop? The first 13 per cent could power as much of the electrical equipment in the house as today's technology does. The 13 per cent from waste heat could provide the power to extract the carbon dioxide which has been absorbed by Aldo Steinfeld's nanostructured gels. Eventually some of the 40 per cent efficient electrical power could be used to turn the carbon dioxide into solar fuels. We will meet two options for doing this that are currently being researched in many laboratories. While we are waiting for this technology, the CPV system could produce as much extra electrical power as today's technology for the same roof area. This could power an electric car for a year via a spare battery. It could also supply the relatively small amount of electrical energy necessary to pump the cooling water and carbon dioxide captured by Aldo's nanostructured gels.

We can have enough power on rooftops with CPV technology. But will we have enough sunlight energy in the UK to produce useful solar fuel? First we must work out how much carbon dioxide is required to produce the solar fuel to power a fuel cell for 8,430 miles. This, you may recall, is the UK domestic average mileage for a year. Professor MacKay, in the book we met in Chapter 8, says that the average new petrol car in the UK generates 0.168 kilograms of carbon dioxide for every kilometre travelled. That is about the weight of a tin of peas for every mile. We really do need solar fuel, and fast.

If you convert the average UK domestic 8,430 miles in a year into kilometres and multiply by Professor MacKay's number you find that 2,266 kilograms of carbon dioxide are generated by the average petrol car travelling the average domestic mileage in a year.

How much energy does it take to extract this much carbon dioxide from the air? Professor MacKay says 'I'd be amazed if the energy cost of carbon capture is ever reduced below 0.55 kWh per kilogram.' He was assuming an efficiency of 35 per cent for the process. As power from the waste heat of CPV electricity generation is at best only 22 per cent efficient we need to up MacKay's figure to 0.875 kWh per kilogram of carbon dioxide. We can now find the amount of energy need to capture the 2,266 kilograms of carbon dioxide by multiplying 2,266 and 0.875 together. By doing so you can find that the energy required to capture the amount of carbon dioxide released by driving the average new UK car for the average yearly mileage is 1,983 kWh.

This energy is less than the 2,110 kWh which today's rooftop systems produce in a year on average in the UK. This is therefore also less than the energy produced in a year from the waste heat of a CPV system, because this process has the same efficiency (0.13) as today's rooftop technology.

My conclusion is that the energy available in the waste heat of a rooftop CPV system is enough to extract the carbon dioxide from the air which the average new car in the UK emits covering the average annual domestic mileage. The challenge is to produce a system like an artificial leaf that can convert this carbon dioxide into a solar fuel, such as methanol. We will look at how research is progressing on two options.

The CPV cells have been manufactured. We will find that the

small concentrator elements that you need for a rooftop system that resembles a PV panel are under development. Furthermore, Aldo's results in extracting carbon dioxide from the air with his nanostructured gel are approaching the number of kWh for each kilogram that I derived from Professor MacKay's value. So, here and now, an engineering company could develop a rooftop CPV system which would enable concerned citizens to extract enough carbon dioxide from the air to offset all the carbon they generate by running the family petrol car. It would also supply electricity for the home, just like today's rooftop technology, and you would have the extra electrical power to charge an electric car battery.

We will find in Chapter 15 that over large areas of our globe a PV system of a certain power will produce at least twice the energy as the same system in the UK. So a CPV system in such regions will be able to capture at least twice as much carbon dioxide from the air as the same system in the UK. These areas include most of the major oil-producing states. They surely will be interested in developing this system which could eventually replace their depleting oil reserves.

Many of these areas are deserts. We will find that, even in a desert, water molecules are present at much higher concentrations than carbon dioxide. Aldo Steinfeld is working on systems that will extract water and carbon dioxide. That, surely, would make this CPV system economically viable in desert regions.

Before examining the two options for converting carbon dioxide extracted from the air to produce solar fuels, I will give a simplified physicist's overview of the very complicated chemistry that nature found to solve this problem. As far as we know, she found the solution by chance. With her example before us, and using the ingenuity of the scientific method rather than the random process of evolution, we can surely find a way to make direct air capture 'economically viable'.

What is photosynthesis?

The best way to approach this extremely complex subject is by describing Joseph Priestley's observations, which were fundamental to understanding photosynthesis. Priestley was one of a number of famous scientists and technologists who lived in Birmingham in the eighteenth century at the start of the Industrial Revolution. I have a soft spot for the city having taken both my degrees there. Priestley's experiments began the study of the biology and chemistry of photosynthesis, even though he had some odd ideas about the physics of burning.

The experiment was simple. Priestley lit a candle and then covered it with a large glass jar that had an airtight seal. The candle burnt for a while, before dying out. What others were later to show more clearly was that burning requires oxygen. The candle had used up all the oxygen inside the glass jar. Priestley then placed a living plant inside the glass jar and relit the candle. This time the candle continued to burn. This was later explained as the plant giving out oxygen, which keeps the candle alight. Priestley made another important observation: when it got dark, this candle went out.

Priestley's experiments showed that photosynthesis requires sunlight and that oxygen is a waste product. Later, others were to show that, in addition to sunlight, a living leaf requires water and, most importantly for this chapter, carbon dioxide from the air. Photosynthesis turns water and carbon dioxide into a carbohydrate, which is the fuel that the leaf needs to grow. Oxygen is the waste product.

In Part I, when I gave you the generation-a-second trip through the last billion years, I mentioned that evolution was happening extremely slowly in the first few months of our equivalent year. In fact, the algae in the sea were doing something crucial for evolution. Oxygen, the waste product of photosynthesis, was building up in the atmosphere to the levels at which life, which developed in the ocean, could move onto the land, hundreds of millions of years later. That occurred around July for the plants and August for the amphibians in our representative year.

How does photosynthesis work?

Photosynthesis is unbelievably complicated. Given that nature appears to have developed the system by chance, I think the description 'miraculous' is, if anything, an understatement. It would require a whole book to explain photosynthesis in simple terms.

Every electron that takes part in capturing carbon dioxide from the atmosphere has had to survive jumps between ten different complex molecular systems. For every carbon dioxide molecule captured from the atmosphere, four electrons have to complete this tough obstacle course. At two separate stages in this chain of electron jumps, a red photon of sunlight has to be absorbed in a complex molecular system to excite just one of the electrons into a higher energy state. Hence, in total, eight red photons have to be absorbed if all four electrons are to complete the course and one carbon dioxide molecule is to be captured.

The two important and very complicated molecular systems that absorb the two photons are known as *photosystem one* (PSI) and *photosystem two* (PSII). Professor Jim Barber at Imperial describes these two photosystems as 'highly efficient molecular photovoltaic nano-machines'. It is as if the social life of Silicon Street has developed to the extent that two separate community centres are needed as venues for children's parties. Each has a party with a Buzz Lightyear carrying four children, one at a time, on his back up to the first-floor playroom. These four children must attend both parties and jump over another eight obstacles outside in Silicon Street each time a carbon dioxide molecule is taken from the air and converted into carbohydrate. And both parties were arranged spontaneously! Nature developed a much more complicated procedure for photosynthesis than a working solar cell.

Here is a second way to appreciate how miraculous it was that nature found a way to manufacture two highly efficient, molecular photovoltaic, nano-machines by chance.

The 42.5 per cent cell, which we met in Chapter 6, could be described in similar terms to Barber's as a 'highly efficient crystalline photovoltaic nano-machine'. It took two decades of dedicated effort by research students and research associates at Imperial, staff at QuantaSol and JDSU, and collaborators at Sheffield University, IMEM Parma, CIP Ipswich and IQE Europe to develop the cell, and the nanostructures, using the scientific method. To produce output power from sunlight with 42.5 per cent efficiency, each electron (and each positron) has to experience a mere *five* jumps between valence and conduction bands, *three* of which are assisted by photons. Developing that cell with the scientific method, a restricted amount of funding and no national solar cell laboratory to help was a challenge.

By contrast, every electron involved in photosynthesis has *ten* jumps between far more complex systems than valence and conduction bands. Every electron also has *two* jumps assisted by photons in extremely complex photosystems. The fact that this whole scheme was found by chance definitely qualifies for the description 'miraculous'.

Finally, here is a third example of the complexity of the way evolution met the challenge of photosynthesis. Despite two centuries of intensive study by many chemists and biologists, the fine detail of the extremely complicated structure of PSII was only completed in 2004. This was achieved by Jim Barber and his team at Imperial.

The molecular system PSII that Jim Barber studied uses the energies of the absorbed photons to split water molecules into oxygen protons and electrons. The protons and electrons can then react with a carbon dioxide molecule and turn it into the carbohydrate the plant needs to grow. PSII first appeared on our planet about 2.5 billion years ago in micro-organisms consisting of single biological cell or a number of biological cells like algae or bacteria. Jim Barber describes this event as the 'big bang of evolution'. For the first time, living organisms had available an inexhaustible supply of protons and electrons from the splitting of water to convert carbon dioxide into the molecules of life. From that moment, living organisms on earth could prosper and diversify on an enormous scale. PSII had established itself, in Barber's words, as the 'engine of life'.

Don't forget that while PSII was achieving this miracle, the waste product, oxygen, was being dumped into the atmosphere. Had this not been happening for more than two billion years, it would not have been possible for the plant life and then animal life to evolve and then spread from sea to land. Then, hundreds of millions of years later, this oxygen was needed when the most advanced form of life started burning the fossilised remains of the plants, algae and bacteria which had decayed deep underground, with potentially disastrous consequences.

What can we learn from photosynthesis?

This very simplified overview of photosynthesis should still enable us to learn some important lessons.

First, if nature found such a complicated way to extract carbon dioxide from the air and turn it into plant fuel *by chance,* application of the scientific method will surely find a simpler and quicker way.

Second, the two approaches to artificial photosynthesis that we will meet have taken inspiration from nature, but mimic it to different extents.

Third, photosynthesis in plants directly produces a solar fuel, a carbohydrate, which the leaf needs to grow. Any artificial photosynthesis approach we discover can produce a solar fuel we can use directly. We will be cutting out the whole time-consuming cycle of plant growth, decay, hundreds of millions of years of oil formation underground, drilling, extraction, refining and transport. Many stages of this cycle result in the production of excess greenhouse gases.

I will now outline two possible approaches to artificial photosynthesis. I expect more options are being studied by biologists and chemists, but I will focus on the approaches relevant to what I believe to be an important application: turning sunlight and carbon dioxide on a domestic rooftop into solar fuel to power the family car.

Biological solar fuels

The first approach starts from the observation that evolution spent two-and-a-half billion years developing very many different types of algae-like micro-organisms. They all use sunlight to turn atmospheric carbon dioxide and water into the fuel on which the micro-organisms thrive. Why reinvent such a productive wheel? Why not find an existing micro-organism that will produce the solar fuel you need and cultivate it in hospitable environments? The challenge then becomes to find an environment in which they live longer and biological catalysts that will speed up the reactions.

One environment in which micro-organisms appear to thrive is inside the very small tunnels inside porous silica; that is, silicon oxide with holes in it. A second promising environment for micro-organisms is inside a few-nanometres-thick (equivalent to several atomic layers) coating of porous plastic or crystalline material.

The silica and the nano-coatings have to be porous so that the water molecules and the carbon dioxide molecules can get in to the micro-organism and the molecules of solar fuel can get out.

The search is now on through the amazing plethora of micro-organisms which evolution has bequeathed us for the micro-organisms to produce solar fuels. Apparently there are examples of micro-organisms that produce methanol molecules, though artificial synthesis of ethanol would be more difficult.

One observation gives me hope that this could still be a promising approach for ethanol. Many micro-organisms produce a form of sugar that plants use as fuel. It is well-known that there are some other micro-organisms that convert sugars to ethanol: they are called yeasts. It seems possible the researchers could find habitats in which one micro-organism turns carbon dioxide into sugar and a second one turns the sugar into alcohol, rather like a two-junction solar cell.

The artificial leaf

Many researchers around the world are looking at different ways to produce an artificial leaf that can generate solar fuels from sunlight. Much of the work can be described as researching one form or another of *photo-electro-chemical cell* (PEC). They are like the cells used for the electrolysis of water that we met in the last chapter, but with one or both of the metal electrodes replaced by semiconductors.

Recall that in electrolysis, electrons are pulled out of the water at one electrode. At the other, electrons are pushed into the water. Pulling and pushing electrons out of a metal should remind you of the Silicon Street party that was my analogy for a semiconductor. Children are directed by an adult out of the door on the first floor of the community centre, which represents the conduction band. The children come back through the door on the ground floor, which represents the valence band, and flop down into an empty chair.

THE FUTURE OF THE SOLAR REVOLUTION

In a PEC the metal electrodes are replaced by semiconductors with the right sort of diversity diode, so that the internal electric field (represented by the adult) points in the appropriate direction. Then photons, which are absorbed in the semiconductor near a surface where electrons are required to *exit*, will raise electrons from the valence band to the conduction band. The field will direct them out into the water. Also, photons absorbed near a surface where electrons are required to *enter* will raise electrons from the valence band to the conduction band leaving behind positrons. These will attract electrons from the water into the valence band.

Semiconductors have another advantage over metals. They have two electron bands (conduction and valence). Also semi-conductors can be found with many different band-gaps (that is the height of the first floor above the ground floor in the Silicon Street community centre). With a semiconductor electrode there are many more opportunities to match the electron energies to the energies they will need in the water for the chemical reactions to take place.

These advantages of the PEC approach compared to electrolysis with metal electrodes were recognised a long time ago. Sadly, there is a big problem where semiconductors and water is concerned. It is not just the well-known dangers associated with mains-connected electrical equipment in the bathroom. Many semi-conductors undergo chemical reactions with water. They are not pushing or pulling electrons as required. Semiconductors are giving or taking atoms, to or from, the water. This happens most strongly if the PEC contains an acid or an alkali. The electrodes could soon dissolve, which is not a commercial proposition. A PEC that produces a solar fuel, for example, might have carbon dioxide in the water. Dissolved carbon dioxide is acidic; the chemical name for soda water is carbonic acid.

My Acknowledgements section thanks people whom I met on visits to the US National Renewable Energy Laboratory (NREL) early in my PV career. On one such visit I was fortunate to meet Art Nozik, an expert in photo-electro-chemistry. He was studying quantum wells for use in PECs to generate hydrogen. He made some important discoveries about quantum wells at that time. I made a mental note to myself that, if I was ever able to show how quantum wells could achieve higher efficiency, I should then look seriously at PECs. Little did I guess that it would take 20 years for our technology to reach 40 per cent efficiency or that the leading PV engineer at JDSU who would buy our technology was someone I met on my first NREL visit.

Though the end of QuantaSol was very sad, particularly for the talented team that lost their jobs, it did mean that I had time to look again at PECs. I did so with the help of Geoff Kelsall and his group in Chemical Engineering at Imperial. It became clear in 2013 that the problem was still to find a semiconductor that was not dissolved in a PEC. By contrast, hydrogen produced by electrolysis with metal electrodes had become more of a commercial proposition in 20 years, though catalysts were still expensive.

There has been a considerable amount of research in recent years that suggests PECs are a promising form of artificial leaf for hydrogen production. In particular, many new catalysts have been produced using the latest nanostructure techniques. They can enhance the reactions at the semiconductor surface where electrons exit, or enter, without themselves changing. Geoff and his team are now working with Amanda Chatten and her group from physics on novel ways to illuminate both semiconductor electrodes. They are testing their ideas on a PEC that will generate hydrogen. If this works, they will then move on to look at methanol production.

How difficult will it be to produce methanol in a PEC from carbon dioxide and photons? The semiconductor electrode that is pulling electrons from the water is acting very much like photosystem PSII, which Jim Barber describes as the 'engine of life'. The photons absorbed by the semiconductor are providing an almost inexhaustible supply of protons (through the membrane) and electrons (through the external circuit) to the other electrode, while liberating oxygen.

At the second semiconductor electrode, the protons (from the membrane) and electrons (from the photons absorbed near the surface of the second semiconductor) could combine to produce hydrogen molecules as happens in electrolysis. More challengingly, the protons and electrons could drive a chemical reaction that turns carbon dioxide molecules dissolved in the water into molecules containing carbon, oxygen and hydrogen atoms. If the right catalyst can be found, each proton and each electron can add one hydrogen atom to a carbon dioxide molecule. When four electrons have been added to the original carbon dioxide molecule, one carbohydrate molecule is produced as in photosynthesis. Producing methanol is more difficult as it requires six electrons for each carbon dioxide molecule.

Despite these difficulties, there have been some encouraging developments in the search for catalysts that will speed up the carbon dioxide to methanol reaction. You can read about them in the references in the Bibliography.

While research continues on the methanol PEC cell, hydrogen, from PECs and from electrolysis using PV electricity, could be used to manufacture solar fuels from atmospheric carbon dioxide indirectly. For example, if carbon dioxide could be removed from the air with sunlight from the rooftop CPV system I described earlier, solar hydrogen could be used to convert this atmospheric carbon dioxide to methanol by conventional industrial processes. I doubt that it would be safe to do this in the garage at home.

It could also be practical to use renewably generated hydrogen, again in an industrial setting, to generate from methanol the more complex hydrocarbons required to fuel heavy transport and aviation. This may not be economic yet in financial terms, but it could be economic in carbon terms.

If either of these two approaches – biological or artificial leaf – can convert atmospheric carbon dioxide to methanol effectively using PV on a domestic rooftop in the UK, this can certainly be done even more effectively and more cheaply in the desert of an oil-rich country, where the same CPV system will produce more than twice the electrical energy.

Why not an 'ethanol economy'?

You may be wondering why I am not pushing an 'ethanol economy' rather than the methanol one. If so, I share your appreciation of ethanol's advantage over methanol. Ethanol from biomass is already a commercial fuel in many countries. Ethanol is not a toxic liquid like methanol. Another important advantage is that ethanol produced from atmospheric carbon dioxide will be pure ethanol. That not only means it is ideal for fuel cells it also has well-known recreational uses, like mixing drinks. Ethanol is a more stable liquid than methanol. What is not to like?

Sadly, my chemist friends tell me that because ethanol is more stable than methanol and because it requires a second carbon dioxide molecule to produce one ethanol molecule, it is harder to synthesise directly. This suggests ethanol produced on domestic rooftops will be a second-generation application for the artificial leaf, or whichever approach wins. It will be some time before a home PV system can be advertised that not only provides electricity but also fuels the car, stores electricity for overnight use in a domestic fuel cell and can turn the soda water in your drinks cabinet to alcohol. At least with methanol, all but the last option will be available.

We can now add another reason why an alien civilisation that discovered $E = hf$ before $E = mc^2$ might not bother to colonise earth. Using $E = hf$, they discovered the artificial leaf and were able to stabilise carbon dioxide levels early in their development of carbon fuelled transport. They had no warming planet to escape from.

Indeed, let me picture a scene, which may be happening at this moment in a remote corner of the cosmos. (OK wise guy! By 'at this moment' I mean the scene will be played out as far forward in time as light takes to reach the remote corner.)

Alien X, relaxing on the balcony, calls out to Alien Y, 'Darling, you're not watching that awful reality TV show again, are you? Come and join me and have some of this lovely fruit cocktail that I've made. It's spiked with our own, home-made, solar ethanol.'

'Just a moment,' replies Alien Y. 'I'm watching *Earth Enders*. It's horribly fascinating. Do you know, the earthlings are fighting yet another war over who owns the decayed remains of 100-million-year-old plants?'

'You mean earth scientists *still* haven't invented the artificial leaf? What is the world coming to! I am *so* glad we've decided to have nothing to do with them.'

How quickly can artificial photosynthesis be developed?

When solar fuel can be generated on rooftops, it will not only give a major boost to the electric car, but it will also provide a very useful form of energy storage. I downplayed the role of energy storage when we met all-solar electricity supplies in Part II. This was for two reasons: first, the Kombikraftwerk demonstration needed only a very small amount of storage and, second, because solar sceptics keep on about how we need to develop storage *before* installing PV and wind. Efficient and cheap energy storage would nevertheless boost the applications of solar technologies. Solar fuel will be the most convenient form of energy storage. We will be able to convert it to electricity in a domestic fuel cell whenever it is needed.

I have no idea which of the two approaches to artificial photosynthesis will be the first to achieve commercial viability, or indeed if some very different approach might come up on the outside to take the prize. I therefore cannot predict how fast it will happen.

SOLAR FUEL

To improve our chances of halting global warming, I very much hope it will be sooner than 2028. In 1979, Mathisen predicted this as the date the step-changing solar/laser/hydrogen cycle would achieve commercial viability.

It all depends how soon we can get governments, funding agencies, scientific societies and the fossil fuel industry to take artificial photosynthesis seriously. In the next chapter, I will outline ways to get the fossil fuel industry to participate in this research. I am sure that the development of artificial photosynthesis would move much faster if the fossil fuel industry were to support it as strongly as their current development preoccupations: research on capturing carbon dioxide at fossil fuel stations (carbon capture and storage) and the search for unconventional (and unpopular) sources of oil and gas such as 'fracking'. When will they learn that nature shows us more effective, more elegant and safer ways to get fuel?

I think it is likely that the development of solar fuels is less challenging than the Manhattan Project, which, with unlimited US government support, succeeded in developing two, completely new, nuclear explosives in under three years. I also believe that solar fuel would come a lot earlier than 2028 if governments and their scientific advisers were to show the same commitment to solar fuels as they do to nuclear fusion research. Press stories in 2012 about the construction of the ITER project at Caderache in the South of France emphasise that 34 nations have committed £13 billion. That is just the construction cost. As I explain in the Bibliography, the main UK research council supports fusion research with more than three times the funding of PV. This is for a project that is not expected to lead to a commercial fusion reactor until after 2050. A much smaller commitment to an international laboratory researching solar fuels would certainly accelerate progress towards artificial photosynthesis and deliver many years earlier.

We must have an electricity supply dominated by renewable generation, and much of our transport running on solar fuel, well before 2050 if we are to avoid the catastrophic effects of global warming. If we achieve this, nuclear fusion will be unnecessary.

One reason for the different government attitude is that, in the UK and USA at least, the military are strong supporters of nuclear fusion research. Given the history of military indifference to solar cells, which I outlined in Part I, it was ironic, but encouraging, to read in 2012 that the US Department of Defense (DoD) has decided to join the research on solar fuels. The DoD is concerned that 3,000 army personnel and contractors were wounded or killed in Iraq and Afghanistan in 2007 alone as a result of attacks on fuel and water resupply convoys. In 2012, the DoD received $1.2 billion for biofuel and solar fuel research and $7 billion for procuring renewable energy. These are large sums by PV standards for every country (apart from China), but small for the DoD.

Now that we have met two approaches to artificial photo-synthesis, I am, at last, ready to outline my solar manifesto. It suggests how we, as individuals and collectively, can influence our governments to speed up the solar revolution.

FOURTEEN

Manifesto for the Solar Revolution

Power to the People

Black Panther slogan

The most successful revolutions have involved both individual actions at the grass-roots level and collective action. One aspect of renewable technology makes me optimistic that the solar revolution can expand fast enough to slow global warming. A wide range of solar technologies can be implemented both at the individual level and on the large scale. Remember, small is beautiful, and also fast.

I provided a number of examples of small and large solar technologies in Part II: rooftop PV and solar farms, vertical-axis wind turbines and offshore wind farms, rooftop solar thermal and concentrating solar power in deserts, ground source heating and deep geothermal energy, small-scale hydropower in rivers and large dams, gas from anaerobic digestion of animal waste on farms and the industrial scale generation of biogas from waste. In the future I hope we will be able to add: CPV on buildings and in deserts, solar fuel from rooftops and on the large scale in oil-rich countries.

What we are discussing really is revolutionary. For more than a century we have assumed that electricity is a commodity that is produced centrally in large generators a long way away. Nowadays people can produce their own electrical power at home. I have called this distributed generation up to now. You may also have seen it described as *micro-generation*.

If the solar revolution is to maintain momentum, gas should increasingly be produced from local agricultural silage, farm waste and landfill. Eventually, I believe, solar fuel will be produced on domestic rooftops. The solar revolution will give new meaning to the 1960s' slogan 'Power to the People'. No wonder some of the strongest solar sceptics are to be found among the senior employees of the gas and electrical generation and distribution companies.

In this chapter, I want to summarise our findings by suggesting some individual actions that we all can take, here and now, to further the solar revolution. I will start with the cheapest and least demanding options. Then I will suggest ways to help the environmental movement counter the influence that the fossil fuel and nuclear lobbies have with government. The next chapter contains suggestions on how the revolution can be exported around the world.

What we can do individually

1. Switch to a renewable electricity provider.
2. Lobby your friends, school, university, bank, church, gym, employer, political party, local authority ... to switch to an all-renewable electricity supplier.
3. Switch to a biomethane supplier as soon as that becomes possible. If not, but you do have a renewable electricity supplier, consider replacing gas cookers and heaters with efficient electrical units and gas central heating with efficient electrical central heating.
4. Join the environmental movement and sign up for the internet campaigning groups that focus on the environment. Help them take on the solar sceptics.

5. Join, buy shares in or set up, your local community-owned renewable power group.
6. Cut down your air and car travel whenever possible. Instead use the internet, car share, street-car rental, public transport or bicycle.
7. Ensure your house is well insulated.
8. Install a solar thermal hot water system or a PV system that trickles the excess electricity into the hot water tank.
9. Install a ground source or air source heat pump.
10. Buy an electric car on your own or with your local community power group.

What we can do collectively

Environmental groups like Greenpeace, Friends of the Earth and the Renewable Energy Association, the new internet campaigning groups, local community-owned power groups, environmental action groups like Frack Off and charities like Pure Leapfrog and Zero Carbon World are all doing an excellent job. Their efforts are particularly impressive considering how much smaller their financial resources are when compared to the fossil fuel lobby. My aim has been to provide information that will help them in their work. Here are some suggestions for areas in which, if they are not already doing so, they might like to focus their campaigns.

Lobby governments to set a date for a 50 gCO2/kWh carbon limit to be imposed soon on all electricity generation

The message in this book is simple. It is practical and cheaper to impose, *here and now,* the 50 gCO_2/kWh carbon limit on electricity generation that the UK Climate Change Committee recommended for 2030. In Part II, we saw that there are no technical or cost reasons why not. Imposing this limit sooner rather than later will have the added benefit of freeing up more subsidies for the renewables.

In the case of the UK government, it will be important to make it clear that this limit does not preclude the 'dash for gas', which they favour. It only rules out a dash for *natural* gas. There are plenty of sources of gas that generate electricity with a carbon footprint below 50 gCO_2/kWh: anaerobic digestion of wastes and many biogas from biomass approaches. We should be campaigning for more effective government stimulation policies for all these options. The cost of this type of electricity will fall as their contribution increases. In contrast, the future price of natural gas from foreign sources is unpredictable, but is certain to rise as resources deplete.

Lobby governments for all-solar electricity and gas supply companies

In countries without genuine all-renewable electricity supply companies or all-renewable gas suppliers, it will be necessary to lobby for legislation and certification schemes to ensure that such companies can trade. They can already do so in the Britain. What is needed in the UK is effective stimulation policies to encourage biomethane production from all wastes including landfill and anaerobic digestion of farm waste, waste food and sustainably farmed biomass.

It needs to be made clear to governments that there is absolutely no need for natural gas produced by fracking for either electricity production or for use in national gas grids. It would be infinitely better for the fight against global warming, and for cheaper energy prices, if, instead of effectively bribing farmers and local authorities to accept fracking, government funds were used to encourage farmers to send their waste for anaerobic digestion.

MANIFESTO FOR THE SOLAR REVOLUTION

Call for effective feed-in tariffs for PV and all other micro-generation

All countries should install feed-in-tariff (FIT) arrangements for PV and distributed generation from other solar technologies, similar to those which have been so successful in Germany. The guaranteed prices for solar electricity should be set at levels that encourage installations to expand at the exponential rate that Germany achieved for PV in the years 2005–2010. As in Denmark and Germany, the incentives should be set at levels that encourage community and corporate investment in renewable systems. Similarly to the German FIT, there should be a pre-agreement with the local solar industries about how the guaranteed prices for electricity should fall as the amount of solar power installed nationally rises. I fear that PV FIT in the UK is being reduced too quickly at the time of writing.

Campaign for effective incentives for solar thermal and heat pumps

Governments should be encouraged to initiate or improve incentives for solar thermal hot water systems and ground and air source heat pumps. Incentives for heat pumps technology should follow the successful example of Sweden.

Co-operate with local authorities and councils

Local authorities like Woking and Kirklees councils have led the way in the UK by initiating their own renewable schemes. They should be held up as good example for other authorities to follow. There is probably much that local authorities and community

277

power groups can learn from similar groups in countries like Denmark and Germany where local power ownership has been established longer.

Call for effective incentives for electric cars and the end to fossil fuel subsidies

Government subsidies relating to electricity generation from fossil fuels and oil and gas exploration should definitely go. The only gas subsidies should be for biomethane and biofuel production from anaerobic digestion, all other forms of waste and environmentally sustainable biomass. There should, however, be subsidies for research and development of the production of solar fuels from atmospheric carbon dioxide with solar power.

There certainly should be no subsidies for exploiting the environmentally damaging and unpopular 'unconventional' fossil sources. There should not be subsidies for implementing the new technology of carbon capture and storage until research and development has clearly demonstrated carbon dioxide emissions compatible with the Committee on Climate Change limit of 50 gCO_2/kWh.

The first step in lobbying any government should be to obtain accurate figures on the subsidies. Be aware governments may try to hide the true comparison of fossil fuel and renewable support by, for example, confusing subsidies, tax breaks and levies.

Naively one might expect that lobbying on this issue would be pushing at an open door in this time of massive cuts in government expenditure. However, the recent increase in fossil fuel tax breaks in the UK suggests that, like many other governments, they are not prepared to stand up to the fossil fuel lobby. I therefore think the fossil fuel companies need to be lobbied directly.

MANIFESTO FOR THE SOLAR REVOLUTION

Try to convince fossil fuel companies that they should be developing solar fuels

We need to convince the fossil fuel companies that it is in their own best interest to produce more biomethane from waste, geothermal heat and make a much greater contribution to research and development of solar fuels. These are the only long-term replacement for the earth's dwindling stock of fossil fuels. Solar fuels are going to happen. This may be sooner rather than later, now that the US Department of Defense is at last interested. It makes economic sense for the fossil fuel companies to be part of this particular revolution. They themselves should be aiming to produce solar fuel in the sunny lands where their depleting, and therefore increasingly expensive, fossil fuel reserves lie.

It is also in the fossil fuel companies' interest to find a more publicly acceptable source of their raw material than the current 'unconventional sources'. The fossil fuel industry is going to come under increasing pressure from governments and the public to stop these policies as the effects of global warming become more acute. Exploiting the unconventional sources can only lead to an acceleration of global warming.

It will not be easy to get the leaders of our fossil fuel industries to appreciate that helping to develop solar fuels is in their long-term interest. I therefore think it important to confront the fossil fuel companies, and the banks that fund their explorations, directly via shareholder campaigns. I am pleased to see that this has already started. Operation Noah is a new group set up to do just that. One of the internet lobbying groups, 350.org, is taking on the European Bank for Reconstruction and Development over their funding of coal plants and fracking. The World Development Movement is challenging banks, pension funds and other finance companies that are funding dirty energy projects.

Such shareholder campaigns will benefit from being able to

accentuate the positive. We are not just saying 'Don't invest in technologies that are environmentally damaging, carbon heavy and unpopular with the general public.' Instead we will be suggesting that they invest in technologies that will not only be popular but have every indication that they will become cheaper. Shareholder resolutions should aim to get the fossil fuel companies to spend more on developing biogas and biofuel production from anaerobic digestion, all other forms of waste and environmentally sustainable biomass. Oil and gas exploration companies should be encouraged to use their expertise to follow the lead of Greenfield Energy by drilling for geothermal energy in the car parks of supermarkets, factories, and commercial and public buildings.

I was involved in some of the shareholder actions that were successful in getting a number of major banks in the UK, Europe and US to disinvest in South Africa at the time of apartheid. Though the shareholder motions only achieved the support of a small fraction of shareholders, a number of banks decided to disinvest. This was partly because of shareholder pressure and boycott campaigns, partly for reasons of public image, but most importantly because they eventually realised that apartheid South Africa was not a good long-term investment. A number of the institutional shareholders in oil companies would surely appreciate that more research and development on a long-term replacement for oil would be a prudent way to protect their investments in the long term.

The Smith School of Enterprise and the Environment at the University of Oxford produced a report in 2013 on what they call fossil fuel 'divestment'. It contains much useful information. The report's conclusion, however, that the impact will be small, misses the point. The shares held by the objectors in the successful apartheid campaign were very much in the minority.

Bodies like the churches, charities, local authorities, universities and pension funds were the main institutional supporters of anti-apartheid shareholder resolutions. In the case of solar fuel campaigns, insurance companies may also wish to join shareholder resolutions. They are well aware of the threat that global warming presents to their balance sheets.

You will note my suggestions do not include peaceful direct action, though I salute the brave solar revolutionaries who do so, whether it be at fracking sites or in the Arctic. It was outrageous that the Greenpeace Arctic 30 ended up in Russian jails. I would rather not suggest any action that I myself would be too scared to perform. It was scary enough speaking, as a single shareholder from the floor, in the hostile environment of the AGM of a large bank in the days of South African apartheid. I see my job as providing information for members of the environmental movement to confront the fossil fuel lobby and the solar sceptics.

One important lesson to be drawn from these brave solar revolutionaries who are undertaking peaceful direct action is that solar revolution has an international dimension. Another lesson from the fate of the Greenpeace Arctic 30 is that some of my suggestions are only applicable in a democracy, possibly one with a reasonably functioning energy market. The future of the solar revolution in the rest of the world deserves the whole of the next chapter to itself.

FIFTEEN

Tomorrow the World

The battle against global warming has received a transformational boost after China, the world's biggest producer of carbon dioxide, proposed to set a cap on its greenhouse gas emissions for the first time.

Tom Bawden, the *Independent*

My manifesto for the solar revolution is not just intended for the people, industries and governments of Europe and North America. This chapter aims to show that the solar technologies have great potential in the rest of the world. Even in northern regions, surprisingly, solar technologies are already taking hold. As the quote above implies, China now holds a special place in the solar revolution, so it deserves a section on its own.

Also in this chapter, I will describe two of my fantasies: a world solar laboratory and the hope that solar could become the 'new nuclear' on the world political scene.

First, take a look at the world map referenced in the Bibliography. It colour codes regions with the same sunlight energy resource. You will see that nearly all of Africa, the Middle East, South Asia and Australia, and around half of Central and Southern America have at least twice the amount of sunlight energy in a year as falls on the UK. In these areas, a PV system with the same *power* will provide at least twice the electrical *energy* in a year as in the UK. From now on, I will refer to such countries as 'sun-rich'.

Sadly, many of these countries appear to be making the same mistakes as much of Europe and North America by burning coal, oil, nuclear and natural gas rather than utilising the solar technologies. This is partly because some, but not all, of them have plentiful fossil fuel resources. There are encouraging signs that some governments in these areas are waking up to the fact that the solar technologies offer them a chance to jump ahead of Europe and North America economically. China is the clearest example. The north and western parts of that country are sun-rich.

Extending the solar revolution to sun-rich countries

Before discussing how the solar revolution could develop faster outside Europe and North America, I will discuss one development in solar co-operation that worries me. The DESERTEC project plans to cover parts of the North African desert with concentrating solar power plants (CSP) and to bring the resulting electric power to Europe by cable.

This seems unnecessary. One of the fastest growing contributions to electricity demand in southern Europe is the use of air conditioners in domestic and office buildings. DESERTEC aims to bring electrical power generated from the sun in North Africa more than 3,000 km or so to satisfy a rising demand in buildings caused by the same sun beating down on the roofs of these buildings. Solar-powered air conditioning is a far more sensible option, which deserves a section on its own later.

Second, how will the local population respond to seeing their land covered with devices that exploit their solar resource? Many in North Africa are already seeking more democratic involvement in the decisions taken by their government. They have seen the oil resources in neighbouring autocracies exploited by, and for, Western counties without the agreement of their populations.

Third, if the local population is to make use of the electricity generated by the large CSP generators in North Africa, they will need to be hooked up to a modern grid system. It is likely to be the urban elite who benefit. In sun-rich countries, the people who could benefit most from economic development live far from any grid. In such villages, stand-alone PV and small-scale wind power can pump water to irrigate the desert, provide the electricity to cool medicines, recharge mobile phones and, in combination with cheap batteries, provide the power to illuminate school homework in the evening, all vital for development.

My alternative suggestion for extending the solar revolution outside Europe and North America is to encourage the governments and international financiers to fund wind turbine and solar panel *factories* in the sun-rich countries.

These factories can produce electricity generators that are suitable for both on-grid and off-grid applications. They can also supply the PV cells to power replica solar factories and self-sustaining greenhouses; both are technologies that deserve sections on their own below.

When the artificial leaf has been developed, PV systems will be able to generate solar fuels with their excess electrical power. It makes much more sense to export energy in the form of solar fuel than electricity. The governments of the oil-rich countries should be particularly keen to support the research and development of solar fuel production, given that their oil resources are depleting. Such countries already have many of the resources necessary to transport the solar fuel. Another interesting coincidence: the most oil-rich countries tend also to be among the most sun-rich.

Relocation or sale of factories to developing countries would also make business sense for Western countries. Sadly, a number of relatively new PV factories in the West are closing because of competition from the newer, larger production lines in China. Relocation would be better than closing. I have no doubt the Chinese will be looking to set up PV factories in sun-rich countries too. We will look at the implications of such a trade later in the chapter, but first I would like to introduce you to some future PV developments that could be particularly suitable for sun-rich countries. Some of them are made practicable by 42.5 per cent efficient CPV cells.

Solar-powered factories

Once a wind turbine or PV cell manufacturing plant has been installed in a southern country, local entrepreneurs should be allowed to replicate the factories. This will increase the rate at which these technologies expand. On a recent trip to Cuba, I was impressed by the way the Cubans have maintained their 1950s' vintage American automobiles for more than half a century. This is despite the technical and economic barriers imposed by the US. Replicating a wind turbine or PV module factory should be well within the capabilities of most sun-rich countries. Replication in this way eventually means more solar panels and cheaper ones, both important in sun-rich countries.

Ideally, the factories in sun-rich countries should be powered by PV or wind. In 2003, Barry Clive, Massimo Mazzer and I made a proposal to the EU to design an easily replicable, solar-powered, solar cell factory for sun-rich countries. Sadly it was not funded.

Solar-powered air conditioning

Air conditioning is one of the fastest growing contributions to electricity demand in many parts of southern Europe. One can see its effect from the daily electricity demand in, for example, California or Italy. In the summer months, the daytime demand does not peak around noon as in the UK or Germany, but in the afternoon when the air conditioning demand is greatest.

Air conditioning in cities is also a major factor in China, as became clear on my one visit to that country. A noticeable feature of Beijing is how many of the Chinese live in flats in tower blocks. I also noted that one does not need a compass to navigate around Beijing. The direction of north is very clear from the tower blocks. The most north facing wall is the only one on which windows have *not* been replaced by electric air-conditioning units. This is a great advertisement for an air-conditioning system powered by the smart windows I described in Chapter 9. Flat dwellers will be able to stop sweating, use less electricity *and* still enjoy the view.

I believe that government policies should encourage the use of solar-powered air conditioning. One approach would be to put a green tax on electric-powered air-conditioning units bought for houses not supplied by renewable electricity companies. The tax could be used to develop smart windows and reduce the price of ground and air source air conditioning.

A group of Italian researchers including my old friend Massimo Mazzer have launched an Italian company Film4Sun that will manufacture easy-to-install PV systems for balconies and smart window shutters. These can provide enough power to run a dehumidifier or the air conditioning in a room. The aim is to provide a stand-alone system that can be fitted by the householder and plugged straight into the mains. It does not require an expensive inverter as it does not deliver its PV power to the grid. Inverters are relatively large and expensive units because the people who run the grid, quite rightly, only accept electric power that meets strict standards. The Film4Sun system has a clever but cheap piece of electronics, which plugs into a household socket. It delivers the PV power to the air conditioning and takes the minimum power necessary from the grid, mainly when the sun doesn't shine. Of course, much of the time that air conditioning is required the sun is shining, as the blocked-out windows of high-rise flats in Beijing testify.

Film4Sun is also working with NGOs and small co-operative farms in Burundi in Africa to adapt and spread its PV technology among low-income populations so they can learn how to manufacture and maintain their own low-cost systems.

The smart windows described in Chapter 9 generate electricity as well as *reducing* air conditioning demand by preventing direct sunlight from entering the building. The direct sunlight is diverted onto 42.5 per cent CPV cells. The cells generate electricity, which can power the air conditioning when it is needed and where it is needed. Every time I see a picture of one of those spectacular glass-covered skyscrapers in an oil-rich, sun-rich country I wince. The air conditioning bill must be massive and sadly provide excuses for their governments to spend their oil profits on nuclear power. The transparent tracking blinds in smart windows power the air conditioning as well as cutting the air conditioning demand.

There are two further advantages of smart windows for the glass façades of these high-rise buildings. Apart from prestige, the main reason for building upward is the high price of land. But some of the floors are taken up by mains-powered air-conditioning units so the cost of office floor-space rises again.

Second, many of the occupants of these prestige buildings are finance companies or banks. They will be happy to pay over the odds for a PV-powered electricity system if it also has battery storage. A big concern for such companies is the loss of computer data following a power cut. They are very keen not to lose the details of your account. The company Solarstructure, which designed two smart windows systems, no longer exists. While it did, it was fun to use commercial jargon to describe the USP (unique selling point) of the company as the UPS (uninterruptable power supply). Ironically, it was the financial crisis of 2008, for which the banks and financial institutions were partly responsible, that finished Solarstructure. When the credit crunch hit, the two multinational engineering companies with which we were collaborating stopped all research and development, including smart windows.

Solar-powered greenhouses

Many sun-rich countries have large amounts of desert and have problems feeding their population. This is even the case in oil-rich states like Saudi Arabia. PV- and wind-powered irrigation is clearly one answer. Solar-powered greenhouses are another.

A greenhouse in a desert can act like Priestley's sealed glass jar. Plants can thrive inside if smart windows control the light, temperature, humidity and airflow. We are working on a type of concentrator at Imperial that can convert sunlight to just the right wavelength at which plants thrive best. The electricity from the PV can be used to extract water from the desert air to supplement the moisture emitted by the crops. Both can be circulated within the greenhouse in a self-sustaining way.

Is water from the desert air a crazy idea? No, it is yet another example supporting Ecclesiastes' view that there is nothing new under the sun. Desert dwellers have known for millennia that *dry, cold wells* can extract water from the desert air. The modern version of this is that water can be extracted from the air with the new technology that is being developed to extract carbon dioxide from the air. Even in the desert, there are considerably more water molecules in the air than carbon dioxide so it is easier to extract. I described one example of Aldo Steinfeld's nanostructured gel system for extracting carbon dioxide from the air in Chapter 13. If we can get this system to extract carbon dioxide with PV power, the same system will work even more effectively on the higher concentration of water in the air. The water extracted by such a system can keep a greenhouse in the desert sustainable.

More ideas for the sun-rich states

I have described some ways in which PV, and high efficiency CPV in particular, could have a revolutionary effect on economic development in sun-rich countries. I hope that readers in these countries will feel inspired to think of other applications, as these technologies get cheaper. I am encouraged by the stories I hear of how the mobile phone has enhanced commercial development in Africa and the Middle East. I feel sure there are entrepreneurs in sun-rich countries already exploiting solar technologies in ways that have not occurred to me.

PV and CPV have even more revolutionary potential than mobile phones in sun-rich countries. They can start contributing more quickly than mobile phones, as they do not require the installation of an expensive network of transmitters and receivers. In fact, PV is already contributing to the mobile phone revolution in many parts of the world. Much of the infrastructure in the mobile networks in sun-rich countries is already powered by PV. There are PV-powered mobile recharging points in regions not on an electricity grid.

I suspect that many autocratic southern governments in sun-rich countries were keen to see mobile phone infrastructure installed because their internal security people need to communicate. They are not so keen on PV because their urban elite prefer central power generation in cities. Ironically, the mobile networks, often powered by PV are now facilitating democracy movements in many of these countries. This is introducing the mass of the population living far from a grid to the benefits of solar power.

CPV will be even more suitable for these off-grid power applications. The sunlight concentrators and three times higher cell efficiency will eventually make CPV cheaper than PV in sun-rich areas. CPV provides more electric power for a given area. Also the CPV system, mounted on a pillar, can track the sun throughout the day, unlike a fixed PV system. Another irony is that GaAs technology, on which the mobile infrastructure and the mobile phones depend, is similar to that in CPV, as we saw in Chapter 6.

The solar revolution is underway in the north

Solar sceptics love to show pictures of snow-covered PV panels. My reply is that, if the photograph was taken on a cold sunny day, such conditions are ideal for generating electricity, once the snow has been scraped off. The colder the PV cell, the more efficiently it generates electrical power.

This point was first made to me by the late Robert Hill, a solar pioneer, at the University of Northumbria in the north of England. Robert was particularly encouraging about my idea of using quantum wells. I will never forget his words: 'Someday all high efficiency cells will have them.' His prediction may yet prove correct. The fact that quantum wells make it possible to optimise for different sunlight conditions also means the performance of triple junction cells can be optimised for different temperatures, including a cold environment.

Robert was showing me the UK's first building-integrated PV system on the walls of the University of Northumbria's computer centre. He proudly reported that on a cold, sunny day in February the system produced more electric power than specified in the design: a good example of the way the efficiency of a solar cell depends on the temperature.

Despite their reputation for having little sunshine, some countries at high latitude are leading the solar revolution. This is partly due to the breadth of technologies available to the solar revolution and their complementary nature as described in Part II. I also like to think it may be due in part to the more democratic nature of their governments. I noted in Part II that both Iceland and Norway generate nearly 100 per cent of their electrical energy from renewable power already, thanks to their geothermal and hydroelectric resources. Norway is also making an important contribution to the solar revolution by being a leading manufacturer of silicon wafers for PV. This process uses a lot of electricity. So production from hydropower in Norway means that the final PV panels have even lower carbon footprints.

The PV resource at high latitudes should not be ignored. The vertical walls of buildings are a potentially great PV resource in northern latitudes. One memorable example of this came one busy Friday afternoon at Imperial College London when I got a phone call from a colleague in the Estates department. His boss had just called him from his office, high on the south side of the administration building. 'How are those guys in physics getting on with their smart blinds? The sun on my computer screen is too bright.' I had to mumble my apologies that we had not managed to get funding. The punch line is: this was a late November afternoon in London. The boss was going to have to blank out the window and turn on the lights to complete his energy efficiency report. The moral is that some vertical walls are a great solar resource in northern latitudes.

I mentioned the PV panels on the east- and west-facing elevations of our home. From late spring to early autumn, these pick up plenty of sunlight because the sun rises in the north-east and sets in the north-west. I am reminded of this when looking out of the north-facing window towards midnight on clear nights in June when there is a persistent, yellow glow above the horizon. This is even more striking in Scotland. This always reminds me that not much further north the sun hardly sets at all at that time of year.

This beneficial effect is already being exploited by Kevin Sharpe of Zero Carbon World as far south as Bath. In Chapter 12, I reported that Kevin gets more mileage from his electric car than expected because he has PV on both the south-east and north-west elevations of the roof. Kevin tells me that over a year the unfavoured north-west orientation generates 80 per cent of the electrical energy on the south-east elevation.

THE FUTURE OF THE SOLAR REVOLUTION

The CPV systems that are starting to be deployed in deserts move slowly round a vertical pillar as they follow the sun across the sky. This tracking keeps the sun focused through lenses onto the small, solar cells. If the system can track through a large angle, a CPV system will capture more direct sunlight energy than a fixed, flat panel of the same power. This is an advantage in the desert. It is even more of an advantage in the north.

You may be concerned that the vertical pillars may not survive in the fierce northern winds. A colleague from the company that deployed one of the first CPV systems in Japan says that one example near the coast has already survived a typhoon.

Of course, PV electrical energy generated in northern latitudes will be even more useful when it can be stored cheaply from summer into winter. Yet again, we can learn from nature. Flowers bloom and animals thrive inside the Arctic Circle despite having no sunlight for months on end. The energy they need to survive over winter is generated by the over-abundance of sunshine they experience in the middle of the year. The energy in sunlight is used by the plants to extract carbon from the air and store it in the bonds holding carbohydrate molecules together. When solar fuel has been developed, it will be an effective way to store excess solar energy in the long term. It will be ideal for conversion to electricity in domestic fuel cells throughout the long, dark northern winter.

Remember, there is another way to store excess solar energy summer to winter, which we met in Chapter 9. Those most democratic of peoples, the Swedes, are already exploiting it. Nature stores excess solar energy a few metres underground. By introducing effective incentives for ground and air source heat pumps, Sweden is well on the way to being the first country whose heating comes entirely from solar sources. So, when a solar sceptic tells you that we must find new ways to store electricity before we can exploit the renewables, or need to get our solar electricity from the deserts in the south, point out that a northern country, Sweden, is already exploiting solar energy storage.

The People's Republic of China

China has recently become pivotal in the solar revolution. The speed at which PV manufacturing took off in China was staggering. When I visited Beijing in 2005, there appeared to be little PV manufacturing, though a number of entrepreneurs were making plans for expansion. By 2010, half the world's PV cells were being produced there.

Sadly China is not a particularly democratic country, so some of the suggestions in the solar manifesto may be inappropriate. The Chinese government is at last responding to the concerns of the many in their population whose lives are blighted by pollution caused by burning coal. Increasingly, their PV panel production is being turned to domestic use and wind-powered electricity installations are also rising fast.

It will be a major step in the history of the solar revolution if, as the quotation at the start of this chapter suggests, the Chinese really do, at last, set a cap on their carbon emissions in 2016. No longer will solar sceptics be able to argue that there is no point in supporting the renewables while the Chinese continue to build so many coal-fired stations. Also, hopefully, there is more chance of worldwide agreement on a carbon cap at the Paris climate change negotiation in 2015. Perhaps we, together, can get governments to agree to announce dates at which they will impose a 50 gCO_2/kWh limit on all new electricity generators. It is a simple, cheap and transparent approach.

I do hope that a damaging trade war between the EU and US on one side and China on the other can be avoided as it could slow the expansion of PV worldwide. It seems to me the EU and US should find alternative ways to give more support to their PV industries. The Chinese certainly are. The Chinese government announced an extra $40 billion investment in PV in 2012. That is a massive sum, a lot more even than the US DoD is committing to renewables.

What will the Chinese do with the $40 billion? My guess is one aim will be to catch up the West in third-generation CPV as they already have with the first- and second-generation PV. A major Chinese conglomerate showed great interest in QuantaSol technology just before the company was sold. Chinese interest in CPV may also be stimulated by their strong interest in light emitting diodes (LEDs). In Chapter 6, we saw that an LED is like a solar cell running backwards. LEDs are grown in very similar reactors to CPV cells. I am aware of reports in the trade press in recent years that Chinese companies have ordered very large numbers of the reactors that produce LED and CPV cells.

One reason for Chinese interest in LEDs is that they have a tradition of covering their public buildings in coloured lights. As we found, LEDs use less electricity than conventional lighting. I expect the Chinese have realised that the lights on their public buildings would have even smaller carbon footprints if they use similar reactors to cover the roofs and walls of these buildings with CPV cells to generate electricity for the lights.

Take a close look at the LED spotlight or torch in your local superstore. It looks rather like a halogen spot-lamp with a bulbous head. This is a silver parabolic reflector, which forms a roughly parallel light beam from the LEDs which are inside the small transparent pillars. There are up to nine LED chips in one spotlight.

If the reflector were replaced by a plastic lens, parallel light from the sun could be focused down onto a high efficiency triple junction cell the size of a single LED. This could make one unit of a CPV system that could be manufactured using the mass production techniques that are already bringing down the costs of LED spotlights. A number of people in Europe, the US and, I expect, China are developing this approach. They include my old friend Geoff Duggan, who patiently taught me about quantum wells during my year at Philips Research Laboratories.

Will Western governments face up to the challenge from China? Should we make use of the lead we currently have in LED and CPV technology to try to keep ahead of the Chinese? Alternatively, given the imminent threat of global warming, should we try to co-operate rather than compete? Here is a suggestion that will stimulate PV in all countries.

An international solar laboratory

The increasing Chinese dominance of PV technology is reminiscent of the situation in which European nuclear physics research found itself after the Second World War. In the countries developing nuclear weapons, applied nuclear research was underway to develop reactors that would generate electricity, as we found in Chapter 4. The more fundamental research into what makes up a proton or a neutron required highly energetic and highly expensive machines. These smash protons into other protons so as to split them into their constituents. Though it turned out that a single low energy neutron could split a massive nucleus, splitting a single proton with another proton requires much higher energy because their natural tendency is to repel each other.

The US had the resources to develop such machines. By contrast, European industry and technology was in a very bad shape. This was why the European nations banded together to form CERN in Geneva so that Europe might compete with the US in what was to become an extremely expensive research activity. A fruitful competition in particle physics existed between Europe and the US for several decades as the energy of the machines rose higher and higher. Then, when a really expensive, higher energy machine was needed to produce the now famous Higgs boson particle, the pendulum of physics swung yet again. A tunnel was dug in Texas, the first magnets were installed and my old friend Tom Dombeck tested them. Then the US decided that their machine would be too expensive. They joined forces with CERN to build a world machine in Geneva.

Working together, the world's physicists achieved their extremely difficult objective. The words 'needle' and 'haystack' are appropriate. The Higgs boson was discovered and the two theorists who predicted it, the Belgian François Englert and the Englishman Peter Higgs, shared the 2013 Nobel Prize in Physics.

The Higgs boson, or God Particle as the press likes to call it, is an infinitely heavier, and very much more elusive, version of the super-hero of our story: the photon. By contrast, our photons are ubiquitous. Nature has provided us with an over-abundance of them. She has already discovered ways they can convert atmospheric carbon dioxide to carbohydrate fuel. World-wide, scientists have made great strides in developing carbon dioxide absorbers and the artificial leaf. In my mind, the technological challenge of using the photons falling on a rooftop to produce solar fuels is less than that of discovering the Higgs boson.

I also believe that the threat of global warming is such that we need to accelerate our efforts to exploit solar technologies and to develop solar fuels. An international organisation, including China, to research and promote solar energy would help. Such a laboratory was first suggested by Joseph Rotblat in 1979, the same year in which Mathisen made the predictions in Chapter 11. Rotblat was the nuclear physicist who left the Manhattan Project as soon as it became clear that Germany was not developing a nuclear weapon. I mentioned that he set up Pugwash, with Einstein's support, to promote international collaboration to reduce the threat of nuclear war. Rotblat and Pugwash received the 1995 Nobel Prize in Peace.

Rotblat made his suggestion in 1979 because he was concerned about the Non-Proliferation Treaty (NPT) and the role of the International Atomic Energy Agency (IAEA), which monitors adherence to the NPT. Under Article IV of the NPT, states with civil nuclear technology are encouraged to co-operate with non-nuclear weapon states in 'the application of nuclear energy for peaceful purposes'. Rotblat feared that such co-operation could lead to the spread of nuclear weapons. Such concerns are still very real today. Iran has signed the NPT and has co-operated with Russia and France under Article IV.

Rotblat felt the treaty should be amended to encourage co-operation in the *most appropriate* technologies for non-nuclear states. If they were sun-rich states, he thought, as I do, that renewable technologies are far more useful than nuclear. He believed a new international body to promote renewable technology would balance the IAEA. The IAEA is best known for its safeguards work, but the promotion of civil nuclear power is part of its remit. Indeed, even in 2010, the IAEA safeguards work, for which they were awarded the 2005 Nobel Prize in Peace, only consumed 38 per cent of the IAEA's budget.

One important contribution by the IAEA has been their support for the International Centre for Theoretical Physics (ICTP) in Trieste, Italy. This institute has been very successful in stimulating the study of all disciplines of theoretical physics in the developing world. It was led for many years by Professor Abdus Salam who received the Nobel Prize for his work on electroweak unification. My last-minute summons to Salam's Nobel press conference was described in Chapter 2.

When I had switched to PV, Abdus told me that he was keen to start a laboratory next to ICTP to encourage experimental researchers from the developing world to study PV. Given his status, an international laboratory promoting the research and exploitation of third-generation PV might have happened. But Abdus, sadly, was not a well man. He died before our plans had progressed very far.

The solar industries do have the International Renewable Energy Agency (IRENA), which was founded in 2009 and which promotes international co-operation in renewable energy. The website indicates that they had 21 employees in 2013. I think IRENA deserves more support from the international community. International co-operation in nuclear energy is being promoted by a large proportion of the 2,300 staff at the IAEA.

What I propose would have a complementary focus to IRENA. A world laboratory would co-ordinate and extend the research and development of solar technologies, in particular of solar fuels and third-generation photovoltaic applications. Like CERN, or the fusion research laboratory ITER, it would be supported by governments, and by national solar laboratories. It would be very much cheaper than CERN or ITER. It should deliver rooftop solar fuel far more quickly than ITER might deliver power.

If the lobbying outlined in my solar manifesto is successful, financial support might be forthcoming from the fossil fuel industry and the governments of oil-rich states. It is likely that large engineering, semiconductor companies and even car manufacturers might be interested in supporting the new laboratory.

Given the urgency of tackling global warming, and to save the lengthy bickering that always accompanies the choice of location for international laboratories, I suggest the new laboratory be based on an existing one. To get things moving fast, I suggest a split site: the German Fraunhofer ISE could host the third-generation PV activities and the US National Renewable Energy Laboratory the solar fuels.

I suggest also, that these two laboratories be named after two internationally respected and far-sighted Nobel Prize-winners, Joseph Rotblat from Europe and Abdus Salam from Asia.

Solar the 'new nuclear'

In the days of the Cold War, East and West competed to sell nuclear reactors to non-nuclear countries on favourable terms. I have argued that it would be more appropriate to the needs of developing countries if Western and Chinese manufacturers were to compete in supplying solar factories. Solar technology might then become a tool of political influence, the 'new nuclear', with the great advantage that it cannot be used for weapons.

In Iran for instance, I suggest that, in return for stopping all uranium enrichment activity, the US and EU should offer to build a solar cell factory and a wind turbine factory, each capable of manufacturing systems producing 1 GW of electrical power a year. This would cost less than a new nuclear reactor. Within ten years Iran could have around 15 GW of new electricity capacity, much more than its nuclear programme will produce in that time. As neither wind nor PV requires fuel, they would give Iran much more energy security than a nuclear programme. Energy security is claimed to be the reason for the Iranian enrichment programme. Hence the leaders of Iran would not lose face with such a compromise.

Remember that in many parts of Iran a PV system of a particular power will produce at least twice the electrical energy in a year of the same system in the UK. Both PV and wind will generate electric power in areas of Iran without a grid. Both are more secure than nuclear against disruption by earthquakes and hostile neighbours.

An even more intractable problem might benefit from solar diplomacy: the long running Israeli–Palestine conflict. The US, and perhaps Europe too, might again offer wind and PV factories, with possibly the addition of solar greenhouse technology, as an incentive for both sides to enter serious negotiations. New elections should be held in Gaza, the West Bank and the refugee camps to mandate Palestinian negotiators. Solar factories would fuel economic development throughout the region. Gaza would no longer be dependent on Israel for electricity.

Crucially, the dispute is fundamentally about the ownership of land, some of which is regarded as sacred on both sides. But who owns the wind, sunlight and atmospheric water above the disputed ground? Solutions to disagreements over the most problematic of the disputed areas could be facilitated by international guarantees that the wind, PV and air from these regions could supply electricity and water to both Israel and Palestine.

My proposals are getting more challenging, so it is time to move on to our final chapter. This will sum up my arguments and speculate on when my predictions might be fulfilled.

SIXTEEN

Wishful Thinking?

> The suggestion that 'the government should introduce legisla-
> tion to make energy companies reduce their use of foreign gas
> and coal and increase the power they get from Britain's wind,
> sun, waves and tides ...' was supported by 85% of respondents.
>
> YouGov (2012)

In this final chapter, I will attempt to update David Mathisen's
prediction that a solar revolution will ensure our civilisation
reaches 2079, the bicentenary of Einstein's birth. The challenge
is to do so without succumbing to the threats we face from the
burning technologies: global warming, fossil fuel depletion or
nuclear disaster. As we have seen, our civilisation is on the cusp.
Will the misapplication of Einstein's $E = mc^2$ finish our civilisation
before the technologies based on $E = hf$ can save us?

The three threats we face are interlinked. Just think about the
following chain of concerns. Most scientists who are not in the
pay of the fossil fuel industries agree that burning fossil fuels
makes a major contribution to global warming. Access to oil was
a factor in the Gulf and Iraq wars. Oil's contribution to future
conflicts is likely to increase as the reserves deplete. Instability in
the Middle East, where much of our oil comes from, is height-
ened by one state, Israel, having nuclear weapons and uncertainty about whether other states like Iran and Saudi Arabia are
seeking them. Any conflict will affect oil supplies, as the latter
two are major oil producers. These geopolitical concerns and

the threat of oil depletion drive governments and oil companies to spend vast amounts on developing technology to exploit the so called 'unconventional' fossil fuel sources: shale oil, tar sands and fracking. These technologies are damaging the environment and have high carbon footprints. If these new fossil fuel resources are burned they will greatly add to the greenhouse gases in our atmosphere; which brings us full circle back to the threat posed by global warming.

The solar technologies give us the chance to break the cycle. The 85 per cent figure in the YouGov survey quoted at the beginning of this chapter is typical of many surveys that show public support for renewable electricity generation. As we found in Chapter 10, our political leaders, their scientific advisers and some of our learned scientific societies think they know better than the bulk of the population. I have argued that this is due to the power of the fossil fuel and nuclear lobbies, with some possible influence from the military.

The YouGov poll suggests that a good proportion of the UK population would support the appeal in the solar manifesto for a 50 gCO_2/kWh limit on new electricity generators here and now. We also have the chance to make our views on solar electricity known to our political leaders by switching to an all-renewable electricity supplier and by participating in the other suggestions in the manifesto.

The arguments in *The Burning Answer* will now be summarised in three open letters and a timeline for the solar revolution. The open letters are intended for the leaders and members of the powerful organisations we still need to influence if the solar revolution is to succeed. They are all busy people and I doubt they have time to read the whole book.

I hope also that these open letters will give environmental and internet campaigning groups some arguments to back up their efforts. They may also give you suggestions for those boxes on internet petitions, where campaigners urge us to personalise our appeals.

To whom it may concern

To the directors, staff and shareholders of EDF

I am writing to you as a concerned Somerset resident. I used to live in Glastonbury. The town will probably lie within the likely exclusion zone should, heaven forbid, a Fukushima-sized accident occur at one of your Hinkley Point reactors.

I am sure we agree about the need to act fast on global warming. The lives of many of your workers at the Hinkley Point site have been affected by the flooding on the Somerset Levels. The increasing frequency of storm surges and rising sea leavels must be a concern to you, given the number of nuclear reactors that you have on the coast.

These events convince me that we need to be installing low carbon electricity generators here and now. I do not feel we can wait until 2023 which I understand is the earliest date your Hinkley Point C reactor will operate. Renewable technologies are contributing to the grid here and now.

I would argue that the renewables have also demonstrated lower life cycle carbon emissions in operation. By contrast there is a wide range of nuclear carbon footprints in the life cycle assessments (LCAs) in the scientific literature. I look forward to seeing a peer-reviewed LCA for your proposed EPR based on construction and performance data of one or other of the two prototypes, when they eventually operate. The LCA should also include explicit calculation of the influence of the uranium concentration of the ore and in-depth analysis of all greenhouse gas emissions during construction, operation, fuel preparation, decommissioning and waste storage.

Your other concern, of course, is the financial viability of your company. I fear that the 'contract for difference' with the British government is a bad deal for UK taxpayers. Given the public's concern about energy prices, the agreement could turn out to be very unpopular. They will eventually realise that it commits taxpayers to funding a wholesale electricity price that is twice what it is today.

I fear the deal might not turn out to be great for EDF shareholders either. As you know, the two numbers that matter in the 'contracts for difference' are hard to predict: the cost of nuclear electricity and the cost of electricity from all other sources in 2023.

There are, however, two factors we can be more certain about. First, the cost of Hinkley Point C is likely to continue to rise. Modifications will have to be made in light of the performance of the first two EPR prototypes, whenever they start operating. In all engineering projects, something is learned from prototypes. As you well know, the two EPRs are very large prototypes involving a unique combination of engineering disciplines. They have also experienced major design modifications, mid-construction, as a result of the Fukushima disaster.

Second, there will be a lot more onshore and offshore wind, and solar PV power on the national grid by 2023. The cost of the electrical power they supply to the grid will have fallen, as it is already doing in Germany.

I assume that as a responsible company you have a plan B ready should the EU regard the 'contract for difference' as a subsidy, or should the price increase even further when one prototype works. Perhaps you have already considered offshore wind and onsite PV? As you know there have been problems with the Atlantic Array offshore wind farm. It was planned for much deeper water than off-shore from Hinkley Point. The main problems with the Atlantic Array appear to be the water depth and the large and very high wind turbines. The latter were not popular with some of the population in North Devon. Smaller turbines in the shallower waters nearer to Hinkley Point should be much easier to build as they are well-established technology. I saw them working in Danish waters many years ago. Smaller turbines in shallower waters are likely to be more popular in Somerset. Many will see them as a safer option than a new nuclear reactor.

You could also cover the Hinkley Point C site with PV. A solar farm like the UK's largest at Wymeswold in Leicestershire would give you at least 0.1 GW of PV power on your 430-acre site. Part could come from the new CPV technology, which will give you the same power from less than half the land area. With 1.5 GW of wind power, as was originally intended for the Atlantic Array, that would give 1.6 GW power as expected for the first EPR, but very much earlier than 2023.

Based on the Atlantic Array cost and the current price of PV, the construction cost of the renewable system would be £4.2 billion rather than the £8 billion for the first Hinkley reactor. As I mentioned earlier, this EPR cost is likely to rise when the two prototypes come on stream. On the other hand, offshore wind prices will fall because of the shallower water near Hinkley Point and as more turbines are installed in British waters. The UK already has 3.3 GW of offshore wind. That is more power than both your proposed reactors may provide in the mid 2020s. I expect that the British government would be prepared to switch your contract for difference to an offshore wind farm.

A 1.6 GW EPR should, in principle, produce more electrical energy than a 1.6 GW wind farm. But the latest Danish offshore wind farms were producing electricity 47 per cent to 49 per cent of the year in 2013. I am sure there is no need to remind you that in 2008 the present generation of British nuclear reactors only produced electricity for 49 percent of the year.

This highlights another important advantage of offshore wind over new nuclear. When the EPR eventually starts, it is very unlikely to achieve design performance immediately. The necessary remedial work would severely detract from the amount of electrical energy the reactor can produce in the first few years. In the case of a wind farm, if the specified performance is not achieved, the offending turbines can be replaced without affecting the working turbines.

WISHFUL THINKING?

In any case, I believe that comparisons between costs of renewable and conventional technologies in terms of the energy produced in a generator's lifetime can be very deceptive. Where supplying the demand on the grid is concerned, it is the power that is important rather than the energy. What matters is the marginal cost of generating the electric power at the time it is delivered to the grid. For a nuclear reactor, the cost of generating that power depends on the cost of running the reactor, the price of the fuel and the financial commitment in disposing of the waste fuel. For offshore wind and PV, the running cost is a lot smaller and the fuel and waste costs non-existent. So the marginal generating costs of renewable power are far lower than nuclear.

My argument is confirmed by experience on the German grid. The cost of PV electricity is still high in terms of the energy that will be generated in a panel's lifetime. But, at the time that the PV power is supplied to the grid, the marginal cost of generating power is lower than all other forms of electricity generation. This can be clearly seen because the peak wholesale price of electrical energy has been falling as the amount of PV power on the German grid has been increasing. The peak price of electrical power on the grid occurs at the time of peak demand. During the day in Germany (and in the UK) this occurs around noon. Not only will the PV power generated by plan B be cheaper than competitors, you will be supplying PV power when it is most needed around the time of peak daytime demand.

I appreciate that your majority shareholder, the French government, will not be happy with plan B. They are looking to your company to provide employment for another of the companies in which they have a majority stake: Areva. In turn, Areva are hoping that EDF will pay the increased costs of the next EPR prototype.

I think plan B may well be popular with many of your staff. It will give them the chance to work on projects that start and deliver much more quickly than the EPR. Areva, themselves, have already wisely decided to diversify. I understand that for many years they have been using their expertise in manufacturing large electrical generating turbines to produce wind power in the North Sea. I feel sure they could also make a big contribution to the offshore wind farm of plan B.

For the committed nuclear engineers at Hinkley, EDF and Areva, there will be many employment opportunities decommissioning existing reactors and searching for a safe and long-lasting solution for nuclear waste. Instead of constructing a new type of reactor, which will raise new waste disposal problems because of the higher burn-up of the fuel, they can focus on the crucial task of finding a secure long-term solution to the existing waste. This is very important given the sophistication of modern terrorists. Both the British and the French governments are committed to the multi-billion pound expenditure necessary.

I find the view from Glastonbury Tor spoilt by two features. The flooding on the levels reminds me how fast global warming is progressing. The small dark specks I see in the distance that are the Hinkley Point A and B stations remind me of two further disturbing images: first, 9/11, and the frightening sophistication of modern terrorism and, second, the picture of the stricken Fukushima reactors. The latter highlights the vulnerability of nuclear stations to the loss of cooling which might result from storm surges and rising sea levels. Rows of tiny windmills in the Bristol Channel and the sun glinting on PV arrays would provide a far more reassuring prospect.

To the leadership, MPs and members of the Labour and Liberal Democrat Parties

I am writing to commend to you a policy for the next election, which, according to the opinion polls, a large majority of the electorate would favour. I believe it would also be popular with many of your MPs and a large proportion of your membership.

WISHFUL THINKING?

You do not need me to tell you what a political hot potato the cost of domestic electricity and gas has become. Also, that the big-six energy supply companies are very unpopular. They blame their large above-inflation price rises on the cost of renewable subsidies, though, as you know, their contribution is a lot smaller than the big-six claim. My policy suggestion will lead to lower electricity and gas prices.

You also do not need me to tell you that, following the wettest winter on record, exceptional flooding and storms with increasing frequency and intensity, the British electorate is crying out for immediate action on global warming.

My policy attacks both problems, immediately and effectively. What I am proposing is that you go for a policy that ensures that the energy industries give the highest priority to the cheaper and lower carbon renewable electricity and biogas sources. The renewables either require no fuel at all or the fuel is produced from cheap indigenous resources. They are the least expensive options for electricity generation.

Opinion polls have repeatedly shown that a large proportion of the electorate favours the renewables. They are clearly aware that they are cheaper than relying on imported gas or uranium. Furthermore, the evidence from Germany, where they have more wind and photovoltaic power than the UK, is that the peak whole-sale price of electrical energy is falling thanks to PV.

There are three simple steps to implementing this policy:

First, announce that you will impose a strict limit on carbon emissions from electricity generation for all new construction after a date to be announced as soon as possible after your election. I suggest you set the target at the limit recommended by DECC's Committee on Climate Change (50 gCO_2/kWh). This will be electorally popular because the wholesale cost of electricity will fall and you will be seen to be taking firm action on global warming. Your government finances would also improve, as there would be no need for the large subsidies and tax breaks for the fossil fuel industries, at least as far as they apply to fossil fuel electricity generation.

Second, announce that if elected you will not build any new nuclear reactors. It is now clear that any electricity they produce will be too expensive. It will not be necessary to make the politically embarrassing admission that your party's earlier nuclear policy was wrong. In government, both your parties made it clear that new nuclear build would only be allowed if subsidies were unnecessary. Whatever the EU may rule, many electors are well aware that a 'contract for difference' is a subsidy. It will commit taxpayers to supporting a price for wholesale nuclear electricity for 40 years from 2023 that is twice today's price.

Third, abolish the large subsidies for natural gas. That will make for an immediate improvement in government finances while leaving plenty over to enhance subsidies for biogas production for the national gas grid and for electricity generation. The new, smaller subsidies should support biogas from sustainable biomass, anaerobic digestion of farm and food wastes, gasification of landfill and other wastes. It will reduce the cost of household gas and of biogas electricity generation. German experience has shown that biogas is ideal for backing large amounts of PV and wind power on the electricity grid. The removal of all subsidies that support unconventional fossil fuel production such as 'fracking' will be particularly popular with the electorate.

Finally, given the unpopularity of the energy and fossil fuel companies, I think it would be a politically astute move for your party to make it clear you do not, and will not, accept donations from companies with a large proportion of their work associated with production or supply of fossil fuels.

To the directors, staff and shareholders of a major fossil fuel company

I am writing to you as your company is ranked in the top 20 of the world's highest greenhouse gas emitters according to Richard Heede of the Climate Accountability Institute.

WISHFUL THINKING?

Like many other scientists, and also many members of the public, I believe we must act quickly and decisively to reduce the amount of greenhouse gases emitted into the atmosphere. I believe that the exploration of any new unconventional and conventional sources of oil and natural gas is unacceptable. This is not just in terms of environmental damage, but also because burning the oil and natural gas from any new source is not compatible with achieving our carbon reduction targets. The words of the International Energy Agency support my case: 'No more than one-third of proven reserves of fossil fuels can be consumed prior to 2050' if we are to avoid the worst consequences of global warming. If burning the reserves you already have could lead to environmental catastrophe, why is your industry exploring *any* new fossil fuel sources?

I will suggest some hydrocarbon sources that have lower carbon emissions, are of more proven yield, more secure and cheaper, in both the short and long term, than unconventional sources. Their exploitation will be much more popular with the public to whom you must sell. For example, your drilling companies could use their expertise to provide supermarkets, and other large companies, with the geothermal energy to run their establishments as Greenfield Energy are already doing.

My second suggestion would be to increase your sourcing of biogas for the electricity and gas grids from sustainable biomass, anaerobic digestion of farm and food wastes, gasification of land-fill and other wastes. This should be for both electricity generation and the domestic gas grid. I expect you are aware of the analysis of electricity generation prices by Mott Macdonald in 2010 that was commissioned by DECC. This points out that the cost of electrical energy generated by biomethane from landfill and sewage waste is cheaper than from natural gas. I interpret their report as saying that electrical energy from anaerobic digestion of farm waste would also be cheaper than natural gas if the subsidies, such as those proposed in support of fracking, were diverted to encourage farmers to use anaerobic digestion.

A third suggestion would be to switch your funding and research and development work on unconventional sources of fossil fuels to research on solar fuels generated by renewable energy from atmospheric carbon dioxide by methods such as the artificial leaf. These are the obvious replacements for fossil fuels and it is in your company's best interest to participate in this research.

For the long-term health of your company you should be investing in these areas rather than exploring unpopular, unconventional sources of uncertain yield. I would be pleased to learn that you are already investing in one or more of these areas. In that case, have you considered stopping all investment in new sources of fossil fuels?

Countries where most of your fossil fuels originate are in general well served with the necessary renewable resources, and also have the financial strength to help support the necessary research and development.

By investing in these areas, I believe you will ensure your company is at the forefront of the development sustainable options for the future of industry as fossil fuel resources decline.

Highlights of the solar revolution

Here is a summary of the stand-out events in the history, so far, of the solar revolution. It gives prominence, as the book does, to my own favourite: photovoltaics. I apologise in advance to the supporters and the workers in the many other parts of the solar cornucopia. They all have a very important part to play in the solar revolution as I hope Chapter 8 made clear.

1865   Maxwell sets the solar revolution in motion by demonstrating mathematically that sunlight consist of electromagnetic waves.

WISHFUL THINKING?

1887   Hertz generates and detects long wavelength electromagnetic waves in his laboratory starting the exploitation of radio waves which led to modern telecommunications.

1900   Planck discovers the formula $E = hf$, which tells us how much energy each colour has in sunlight.

1905   Einstein explains the meaning of $E = hf$. Sunlight consists of very small quanta of energy, later called photons. This starts the quantum revolution.

1909   The Nobel Prize in Physics is awarded to Marconi for wireless communication and Braun for his rectifier. Much later Braun's invention was shown to be made of a semiconductor.

1910   Bohr uses Einstein's idea to explain how electrons jump between orbits in atoms.

1922   De Broglie explains Bohr's idea that electrons occupy certain orbits by proposing electrons are waves and particles.

1925   Pauli's exclusion principle explains the importance of full electron orbits.

1926   Schrödinger's wave equation predicts the shapes of electrons in their orbits.

1928   Dirac predicts that electrons are spinning. This means that Pauli's principle gives the exact number of electrons in a full orbit.

1931   Wilson publishes his theory of semiconductors.

1947   Bardeen, Brattain and Shockley start the semiconductor revolution by demonstrating a transistor that involves electron conduction and also positive conduction in a nearly full valence band.

1954   Chapin, Fuller and Pearson demonstrate a 6 per cent efficient silicon solar cell.

1973   Solar Power Corporation reduce the price of silicon PV panels fivefold in two years by using crystals rejected by the silicon chip industry.

1978   Handheld calculators appear on the market powered by amorphous silicon thin film solar cells.

THE FUTURE OF THE SOLAR REVOLUTION

1990   The Intergovernmental Panel on Climate Change (IPCC) warns that greenhouse gas emissions are contributing to global warming.

1994   The National Renewable Energy Laboratory (NREL) produces a GaAs-based, two-junction cell with efficiency higher than any single junction PV cell. Japan launches a 70,000-roof PV programme.

1998   Germany initiates a 100,000-roof programme.

2000   The feed-in-tariff (FIT) is introduced in Germany for most forms of renewable energy.

2002   More than 90 per cent of all new homes in Sweden are equipped with a heat pump.

2004   The price of PV panels stops falling. The expanding PV industry is using as much silicon as the semiconductor industry is rejecting.

2007   Spectrolab manufacture a 40 per cent efficient triple junction cell. IPCC predict how fast the size of the Arctic ice cap will shrink.

2009   The fall in PV panel price resumes as the second-generation thin film producer First Solar becomes first company to produce 1 GW of PV panels in a year. New Chinese silicon manufacturers also approach 1 GW per year using new silicon supply chains.

2010   The German Federal Environmental Agency sets a target date of 2050 by which time Germany will have a 100 per cent renewable electricity supply. Half the solar panels manufactured in the year are produced in China.

2011   In February the Danish government announces that the country will be fossil fuel independent by 2050. The Fukushima disaster heralds the end of the nuclear renaissance. Germany, Italy and Switzerland confirm their non-nuclear futures. The peak price of electricity on the German grid falls below the night-time cost for the first time on 2 June.

2012   The National Snow and Ice Data Centre reports that the Arctic ice cap has shrunk to the size that the IPCC predicted in 2007 would not be achieved until 2065. In the early afternoon of 2 May the cost of electricity on the southern Italian grid falls to zero. Wind and PV regularly contribute over 30 per cent of the power on the German grid without problems. In *Nature Materials*, Barnham, Mazzer and Knorr, propose a moratorium on all new electricity generation apart from the renewables on the basis of the actual performance of solar generation on German and Italian grids. The Chinese government amends its 2011–2015 Five-Year Plan to add a $40 billion investment in PV. The US Department of Defense announces a $1.2 billion investment in biofuels and solar technologies plus a $7 billion procurement contract for renewable power.

2013   China proposes to set a cap on its greenhouse gas emissions for the first time. JDSU announces manufacture of 42.5 per cent efficient QuantaSol CPV cells. In May carbon dioxide in the earth's atmosphere passes the 400 parts per million level for the first time since the start of direct measurements. By mid-year the UK has installed 3.3 GW of offshore wind power, more than twice the power of the proposed new nuclear reactor whose scheduled date for starting slips to 2023.

2014   *The Burning Answer* is published, setting out a manifesto for the solar revolution with suggestions for individual and collective actions.

2015-2020

My wish list for the second half of this decade starts with hopes for agreement on tackling global warming at the Paris Climate Conference in 2015. We cannot afford it to fail like the UN conference in Copenhagen in 2009. The cheapest, most effective and most transparent action would be to agree that each country will set dates for implementing a limit of 50 gCO_2/kWh on all new electricity generation. Agreement to replace fossil fuel subsidies by renewable incentives should also be on the agenda as well as setting up an international solar energy laboratory.

What do I hope to see by 2020? I have predicted that if the UK follows the lead Germany achieved for PV 2005–2010, by 2020 Britain will have all the PV power it needs for a Kombikraftwerk, all-renewable scenario. If it similarly follows the German experience in implementing onshore wind power and sticks to its own plans for offshore wind power, its wind installations will be over half the all-renewable scenario by 2023. By using the fossil fuel subsidies to stimulate biogas, geothermal, hydropower, wave and tidal power generation, the rest of the UK may not be that far behind Scotland which aims to achieve an all-renewable electricity supply target by 2020. I hope that the progress the UK is making towards this target will finally convince the government that new nuclear power is unnecessary.

I hope that by 2020 you will have the chance to buy CPV units containing QuantaSol style triple junction technology for your rooftop. These could supply electricity for your household use, your electric car and also provide hot water. I also hope this technology will be providing electricity from the solar blinds inside double glazing. These will reduce air conditioning requirements in office buildings, flats and houses as well as generating electric power.

I think there is every chance that well before 2020 hydrogen production from the renewables will be shown to be commercially competitive. The production of solar fuels like methanol from carbon dioxide in the air using PV is much more of a challenge. However, I am optimistic that there will be a practical demonstration of the principle by 2020. By that date, methanol fuel cell hybrid electric cars should have entered production.

WISHFUL THINKING?

2020-2028

If we reach 2020 with a large proportion of my wishes for the solar revolution fulfilled, we must have a good chance that, well before 2028, we will achieve a solar fuel breakthrough as Mathisen foretold, though not exactly in the same form. We could have a replacement for petrol, derived by CPV from carbon dioxide in the air in a system suitable for installation on a domestic rooftop. During this decade, I would hope that the national and international solar laboratories would move on to develop solar fuels that are more suitable for powering aircraft, heavy transport and industry.

If we do, I predict that the three existential threats – global warming, oil depletion and nuclear disaster – will have eased and our civilisation will have an excellent chance of making it through to the bicentennial of Einstein's birth in 2079. Thanks will be due to many scientists and technologists throughout the world. Thanks will also be due to very many citizens, including readers of this book, whose efforts will have persuaded electricity companies, fossil fuel companies and politicians that solar technologies are the answer to burning. You will have been proved right.

Acknowledgements

I am particularly indebted to my wife, the poet Claire Crowther, who not only inspired me to write this book but has read and re-read every chapter countless times. If non-scientists find *The Burning Answer* readable it will be thanks to her suggestions. I am also grateful for her epigram for Chapter 7.

We are both extremely grateful for the support we receive from our daughters: Cleo, Esme, Heli and Nicole. Cleo and Nicole created the beautiful artwork for the dedication to their mother that has been immortalised in her headstone. Many thanks to Cleo for the photo on the jacket.

Our daughters have given us six fantastic grandchildren: Christy, Eliot, Jude, Mirna, Olan and Rowan. Thanks to them I have become aware of the obvious and the unexpected links between children's entertainment and quantum mechanics. They have inspired me. This book is my attempt to ensure they will have a sustainable future.

I would like to thank Bea Hemming of Orion Books for her encouragement and for her suggestions for making the book readable for non-scientists, and the team at Orion for their hard and very efficient work. Particular thanks are due to Holly Harley for help with the proofs, and Jennifer Kerslake for her encouragement and advice on the paperback edition. I would also like to thank Andrew Barber for suggesting Orion.

My attempts to bring the technology down to a popular level, without misrepresenting the physics, have been greatly helped by Gron Jones, my oldest friend from particle physics days. His suggestions, based on his experience of explaining physics to non-scientists, have been invaluable. I would like to thank my closest friend in photovoltaics, Massimo Mazzer, for his support over the years, in particular through the trauma of setting up two photovoltaic companies. He and his colleagues at IMEM Parma in Italy have always been most helpful and friendly on my many visits.

I have been extremely fortunate many times in my research career. For this book, my most important stroke of luck was the opportunity to change the direction of my research, mid-career, more significantly than is usual in one academic lifetime. In 1986 I switched from particle physics at Imperial to the semiconductor group started by the late Tony Stradling.

To prepare for the switch I spent an extremely stimulating sabbatical year at Philips Research Laboratories (PRL) in Redhill. PRL was then a world-leading laboratory in the exploitation of quantum semiconductor structures. I am most grateful to the Royal Society and the Science and Engineering Research Council for an industrial fellowship, to Bruce Joyce for introducing me to Philips, and to John Walling and John Orton for accepting an experimental particle physicist into the new world of commercial semiconductor research.

I have many times referred to John Orton's two excellent histories of semiconductors which I have found invaluable. I am also indebted to the patience so many of John's group showed in explaining semiconductor physics to a novice. I am grateful to Peter Blood, Phil Dawson, Tom Foxon, Jeff Harris and David Lacklison. I must give particular thanks to Geoff Duggan (now with Full Sun Photovoltaics): together we wrote the first paper on the use of quantum wells in solar cells. I also learnt a lot about the vicissitudes of commercial semiconductor research when Philips shut down this world-leading research effort a few years later. However, fortunately for me, most of the team obtained research positions in the UK. I therefore had convenient sources of expertise

ACKNOWLEDGEMENTS

to tap, and excellent collaborators to work with, when I started solar cell research at Imperial.

I felt confident to comment on the future of the solar revolution in Part III thanks to my experiences at PRL and while building up the Quantum Photovoltaic (QPV) group at Imperial. The successes of our quantum well solar cell, which included two efficiency world records, resulted from the dedicated efforts of a large number of undergraduate project students, PhD students and research associates over two decades. I would like to thank them all. I must mention the contributions of the co-founder of the group, Jenny Nelson, the co-inventors of the strain-balanced quantum well cell, Ian Ballard, James Connolly, Carsten Rohr and the two who now lead the QPV group, Amanda Chatten and Ned Ekins-Daukes.

Many collaborators helped the QPV group. Particularly significant were the contributions of Geoff Hill and John Roberts and their co-workers at the EPSRC National III-V Facility in Sheffield. The QPV group, or myself personally, received support from: Paul Alivisatos, Neal Anderson, Alex Freundlich, Art Nozik, Jerry Olsen, Rob Stevens and Jan-Gustav Werthen (USA), David Faiman (Israel), Laurentiu Fara (Romania), Bis Ghosh (India), Martin Green (Australia), Antonio Luque and Antonio Marti (Spain), Craig MacFarlane, Gareth Parry, Chris Phillips, Paul Stavrinou and Mark Whitehouse (UK), Rudi Morf (Switzerland), Yoshi Okada, Akihiro Hashimoto and Masafumi Yamaguchi (Japan), Christian Verie (France).

QPV received financial support from the Ashden Trust, British Council, Engineering and Physical Science Research Council, the European Union Framework and Marie Curie programmes and Tata Ltd. Particular thanks must go to Jerry Leggett who was instrumental in providing, through the Greenpeace Trust, the financial support for Jenny Nelson which enabled us to start the research when more conventional funders were sceptical.

Massimo Mazzer, John Roberts and I founded QuantaSol in 2007. Thanks to the efforts of a talented and dedicated young

ACKNOWLEDGEMENTS

team, a novel form of triple-junction concentrator technology was developed which achieved 40 per cent efficiency within three years. It was tragic to see so many talented young people lose their jobs when the company was sold to JDSU in 2011. I particularly wish to acknowledge the pioneers: our first CEO, Kevin Arthur, and our enabling administrator Sue Sparkes. In addition I would like to thank Sean Amos, Jon Burnie, Dave Bushnell, Alison Dobbin, Virginie Drouot, Matt Lumb, Kan-Hua Lee, Alex Metcalfe and Chris Shannon. I must also thank Imperial Innovations and the Low Carbon Accelerator for their financial backing of QuantaSol, even though we disagreed over the eventual sale.

Particular mention should go to the four who joined JDSU to show them how to use QuantaSol technology: Gianluca Bacchin, Ben Browne, Victoria Rees, and Tom Tibbits. Within a year they had not only upped the cell efficiency to 41.5 per cent but also, as I had long advocated, added John Robert's quantum wells in the top cell to achieve a production cell world-record performance of 42.5 per cent efficiency. Despite these achievements, they too lost their jobs when JDSU stopped all third-generation research and development in the light of the fall in the price of first- and second-generation photovoltaics.

Many people have given me a helpful background on photovoltaics and related applications. I am most grateful to Kenji Araki (on CPV), Marie Austenaa (on mobile phones), Geoff Cardnell, Keith Jackson and John Warman (on electric heating), Nick Carpenter (5D Group), Barry Clive and Ralph Hudson (on concentrators), Dave Eastwood (on computers), Ed Gill and Chris Welby (Good Energy), Ralf Gottschalg, Ben Kluftinger and Kaspar Knorr (on Germany), John Hassard (on tidal power), Arnulf Jaeger-Waldau (on air conditioning), Zhaohui Ma, Nicoletta Marigo and Wang Sicheng (on China), David Place (EvoEnergy), Mohammed Redwan (on Gaza), Greg Rhymes (Your Power), Julio Rimada (on Cuba), Kevin Sharpe (Zero Carbon World), Jonathan Wright (Wright Instruments).

ACKNOWLEDGEMENTS

Where technologies other than photovoltaics are concerned, I have been highly dependent on help from experts. David Lowry has been particularly helpful on all matters nuclear and in researching parliamentary energy issues. Ian Fairlie has helped with issues of nuclear radiation. Support has also come from many members of the Nuclear Consultation Group, led by Paul Dorfman and Scientists for Global Responsibility, led by Stuart Parkinson.

Many have commented on part of the text for which I am most grateful. Three are from my own department, Physics, at Imperial: Mike Rowan-Robinson who advised on the astrophysics and Joanna Haigh and Ralph Toumi who helped me with wind-power. Other experts include John Baldwin (CNG Services), Nick Finding (J V Energen) and Ciaran Burns (REAL) who were helpful on biogas. Mads Bang (Serenergy) gave good advice on fuel cells. Ryan Law (Geothermal Engineering) and Dmitriy Zaynulin (Greenfield Energy) gave very useful help on geothermal energy.

Electric cars, photosynthesis and solar fuel were subjects I knew little about until I started Part III of this book. So I am particularly grateful for the help I have received from experts in chemistry and chemical engineering: Michele Aresta, Jim Barber, Franca Bigi, James Durrant, Peter Edwards, Kisuk Kang, Aldo Steinfeld and Bao-Lian Su. In particular I have learnt about the photo-electrolysis of solar fuels from the ongoing collaboration between Amanda Chatten and Jose Videira in Physics with Geoff Kelsall, Anna Hankin and Chin Kin Ong of Chemical Engineering at Imperial.

Many others have provided suggestions or helpful information for the book: Peter Andrews, Martin Bacon, Frank Barnaby, Robert Barnham, Roger Bentley, Shaun Bernie, Neil Crumpton, Steve Cussell, Carolyn Dale, Leon di Marco, Rob Edwards, Graham Ford, Phil Harper, David Hart, David Haslam, Colin Hines, Paul Ingram, Keith Jackson, John Large, Peter Macfadyen, Helen Moore, David Morrison, Alan Nix, Mario Pagliaro, Hugh Riddell, the late Dave Rogers, Steve Thomas, Dave Toke and Greg Watt.

ACKNOWLEDGEMENTS

While thanking all my helpers, I take full responsibility for any errors that remain in the text.

Finally, I hope I will be able to thank you, the reader. The solar revolution, like the semiconductor revolution, is one in which we can all participate.

Bibliography

Introduction

This bibliography provides the references for the original articles and books from which I have quoted data or summarised arguments. I also suggest further reading for those scientists and non-scientists whose interest may have been stimulated in the technology or the political issues. Another use has been to include points that would have been added as footnotes in an academic text.

Stephen Hawking's thought-provoking conjecture about alien civilisations was made on his Channel 4 television programme [1]. The book that explains to non-scientific readers the meaning of Einstein's famous equation $E = mc^2$ was written by Brian Cox and Jeff Forshaw [2]. Bill Bryson's book [3] covers an amazingly wide range of science issues in a clear and engaging style. It is also a book for non-specialists.

Of the three existential threats to our civilisation – global warming, oil depletion and nuclear disaster – the second is the one least discussed in this book. This is mainly because I know least about it. The very real threats from oil depletion are extremely well covered in two books by my old friend Jeremy Leggett [4, 5].

The typical example of the way the energy debate is bedevilled by confusion between energy and power was quoted from the *Wall Street Journal* [6].

The spare-time study of plutonium production in the British civil nuclear reactors that I made with colleagues from Scientists Against Nuclear Arms (now Scientists for Global Responsibility) was published as a *Nature* commentary article [7]. Chapter 4 contains the references to the subsequent information released by the British government, which confirmed the accuracy of our calculations.

The leading US semiconductor company JDSU announced at the CPV9 conference in Miyazaki, Japan in April 2013 that they had manufactured 42.5 per cent efficient triple junction cells with the QuantaSol quantum well technology in both the top and middle subcells [8]. The sunlight to electrical power conversion efficiency of these concentrator cells is three times that of a typical 14 per cent efficient silicon panel on rooftops today. At the same conference, the average efficiency of the cells *manufactured* by the rival company that holds the world record (44 per cent) for one *research* cell was reported to be lower than 42.5 per cent [9].

1. Stephen Hawking, *Stephen Hawking's Universe*, Part I, 'Aliens', Channel 4, 18 September 2010.

2. Brian Cox and Jeff Forshaw, *Why Does E = Mc²? (and why should we care?)*, Da Capo Press Inc. (2009).

3. Bill Bryson, *A Short History of Nearly Everything*, Doubleday (2003).

4. Jeremy Leggett, *Half Gone: Oil, Gas, Hot Air and the Global Energy Crisis*, Portobello Books (2005).

5. Jeremy Leggett, *The Energy of Nations: Risk Blindness and the Road to Renaissance*, Routledge (2014).

6. Ted Nordhaus and Michael Shellenberger, 'Going Green? Then Go Nuclear', *Wall Street Journal*, 22 May 2013, http://online.wsj.com/news/articles/SB10001424127887323716304578482663491426312, accessed 10 December 2013.

7. K.W.J. Barnham, D. Hart, J. Nelson and R.A. Stevens, 'Production and destination of British civil plutonium', *Nature*, 317, 213 (1985).

8. Ben Browne et al., 'Triple-Junction Quantum-Well Solar Cells in Commercial Production'. JDSU presentation at 9th International Conference on Concentrator Photovoltaic Systems (CPV9), Miyazaki Japan, April 2013.

9. Jeff Allen et al., '44%-Efficiency Triple-Junction Solar Cells'. Solar Junction presentation at 9th International Conference on Concentrator Photovoltaic Systems (CPV9), Miyazaki Japan, April 2013.

BIBLIOGRAPHY

1. We are Stardust

'Woodstock' words and music by Joni Mitchell, © 1969 Crazy Crow Music. All Rights Administered by Sony/ATV Music Publishing, 8 Music Square West, Nashville, TN 37203. All rights reserved. Used by permission of Alfred Music.

The information for my brief history of baryosynthesis, the formation of the solar system and the lifetime of our sun was taken from the undergraduate nuclear physics textbook by Kenneth Krane [1]. There is a clear description of nucleosynthesis, suitable for non-specialists, in the Cosmology chapter of *Galileo's Finger* by Peter Atkins [2]. He also suggests some useful references for further reading. I took the limit for the amount of plutonium that a radiation worker could safely inhale from The Royal Commission on Environmental Pollution 6th Report [3].

The dates for my generation-a-second overview of the past billion years were taken from two sources: *Energy: Engine of Evolution* by Frank Niele [4] and a book on genetic evolution by Richard Dawkins [5]. The latter is a fascinating account of how evolution gave us such an incredible diversity and complexity of life forms during the billion years that we skate through at a generation a second. Dawkins also has an interesting discussion about the controversy over how long the underwater Cambrian explosion lasted in the June of our representative year. The Niele book favours three days rather than three weeks.

I first heard the suggestion that we need to return to the situation where the sun is our primary source of energy in a presentation by Richard King [6] though others may have made the point earlier.

The nuclear disaster at Fukushima in 2011 and its impact are described in book written for non-specialists by David Elliott [7].

The crucial observation that the energy in the sunlight falling on the earth in *one hour* is more than enough to supply all the energy demands of humans for *one year* was first made, as far as I am aware, by Nathan Lewis in the journal *Science* [8].

1. Kenneth S. Krane, *Introductory Nuclear Physics*, John Wiley and Sons (1988).

BIBLIOGRAPHY

2. Peter Atkins, *Galileo's Finger: The Ten Great Ideas of Science*, Oxford University Press (2003).

3. The Royal Commission on Environmental Pollution 6th Report, *Nuclear Power and the Environment*, Cmnd. 6618, Her Majesty's Stationery Office, London (1976).

4. Frank Niele, *Energy: Engine of Evolution*, Elsevier B.V., Amsterdam (2005).

5. Richard Dawkins, *The Ancestor's Tale: A Pilgrimage to the Dawn of Life*, Weidenfeld & Nicolson (2004).

6. Richard J. King, 3rd World Conference on Photovoltaic Energy Conversion, Osaka, Japan (2003), 2527.

7. David Elliott, *Fukushima: Impacts and Implications,* Palgrave Macmillan (2013).

8. Nathan Lewis, 'Toward Cost-Effective Solar Energy Use', *Science*, 315, 798 (2007).

2. What is Light?

In the nineteenth century a number of physicists and technologists transformed the ways we generate and use energy. The cultural and economic background to their discoveries, and the rivalries between the protagonists, are described in a book by Crosbie Smith [1]. Details of the lives of the main characters can be found in many sources including the *Oxford Dictionary of Science* and the two-volume *Biographical Dictionary of Science* [2, 3]. The latter has suggestions for further reading.

Crosbie Smith's book has much to say about Maxwell, but the details I quote here, including the letter to his cousin, are mainly taken from a biography by Everitt [4]. In order for you to appreciate the significance of Maxwell's revolutionary discovery, I admit that I may have occasionally placed physics understanding ahead of accurate historical sequence.

If the discussion of electric and magnetic fields is new, you might find it helpful to look at a book written at the school science level [5]. You can see the effects of scattering iron filings over a bar magnet in a picture on the internet [6]. Note that many of the iron filings have fallen in elliptical loops. Ignore the filings that have fallen on the magnet itself.

BIBLIOGRAPHY

The circular loop of magnetic field around a wire carrying a current (the red dot) is clearer in a second picture on the web [7].

I recommend you try the interactive demonstration of a magnetic field around a current carrying wire that can also be found on the internet [8]. It shows an electric current in a wire producing a loop of magnetic field around the wire. You can reverse the direction of the electric current. The magnets then show that the field points round the loop in the opposite direction.

Maxwell's equations are mathematically very complicated. This is because they describe how electric and magnetic fields vary not only with time but also in the three spatial directions. They are therefore usually written in the elegant mathematical shorthand known as vector calculus. Textbooks that discuss his equations are usually written assuming the readers are undergraduates who have taken an advanced calculus course. My attempt to describe them in words is, as far as I am aware, the only non-mathematical explanation that has been published. Sadly, some familiarity with the mathematical symbols of vector calculus is assumed even in Feynman's undergraduate lectures [9]. But the explanations that go with Feynman's mathematics are, as always, both clarifying and stimulating.

The full reference for Stephen Hawking's book is given below [10]. Peter Atkins' book is referenced in Chapter 1.

The race to produce the first street lighting is described on the engineering timelines website [11].

A book written for non-specialists about electroweak unification, which emphasises the impressive experiments that confirmed the theoretical ideas, was written by an old friend of mine from Birmingham University [12]. Peter Watkins participated in one of the experiments.

1. Crosbie Smith, *The Science of Energy: A Cultural History of Energy Physics in Victorian Britain*, University of Chicago Press (1998).
2. *Oxford Dictionary of Scientists*, Oxford University Press (1999).
3. Roy Porter and Marilyn Ogilvie, *The Biographical Dictionary of Scientists*, Vols I & II, Oxford University Press, 3rd edition (2000).
4. C.W.F. Everitt, *James Clerk Maxwell, Physicist and Natural Philosopher*, Charles Scribner's Sons (1975).
5. Gerard Cheshire, *Electricity and Magnetism (Science Essentials – Physics)*, Evans Brothers Ltd (2010).

BIBLIOGRAPHY

6. Wikimedia commons, 'File:Magnet0873.png' http://commons.wikimedia.org/wiki/File:Magnet0873.png, accessed 5 February 2014.

7. Wikibooks, 'GCSE Science/Magnetic effects of a current' http://en.wikibooks.org/wiki/GCSE_Science/Magnetic_effects_of_a_current, accessed 5 February 2014.

8. GCSE Physics, 'Electromagnetism', http://www.gcse.com/energy/electromagnetism.htm, accessed 6 January 2014.

9. Richard P. Feynman, Robert B. Leighton and Matthew Sands, *The Feynman Lectures on Physics*, Volume II, Addison-Wesley (1965).

10. Stephen Hawking, *A Brief History of Time*, Bantam Press (1988).

11. Engineering Timelines, 'Godalming Power station', http://www.engineering-timelines.com/scripts/engineeringItem.asp?id=744, accessed 6 January 2014.

12. Peter Watkins, *The Story of the W and Z*, Cambridge University Press (1986).

3. The Quantum Revolution

The Bohr quotation is beloved by lecturers giving an introductory course in quantum mechanics. There are many versions; this one is taken from reference 1.

My account of Marconi's dream on the Cornish cliff-top is fictional. Gavin Weightman has written an interesting biography of Marconi and the challenges he and others faced in establishing wireless transmission [2].

Most of the physicists and chemists who made important contributions to the quantum revolution were awarded Nobel Prizes. The official website of the Nobel Foundation has interesting biographies of the physics [3] and the chemistry winners. If you click on the laureate's name you can read the biographical details and their lecture.

As I have mentioned, there are many excellent popular books on the history and philosophical implications of quantum mechanics. John Gribbin's book [4] is one of the few that I have read that connect with the practical applications like semiconductors. Though he mentions LEDs, solar cells are not discussed.

The physics details of the ultraviolet and the infrared catastrophes,

BIBLIOGRAPHY

their explanation in terms of quantum ideas and the early applications of quantum theory to atomic structure can be found in all undergraduate physics textbooks. My own preference is that by Young and Freedman [5].

The quotation which shows Pauli's rather negative view of semiconductors can be found in the second reference of Chapter 5. It is ironic as Pauli's ideas about full electron orbits turned out to be fundamental to the discovery of electric currents due to positive charges. This made the semiconductor revolution possible.

I have gone into the question of why full electron orbits are stable because it is fundamental to quantum bonding. Hence it is at the basis of the chemistry of life, the energy in our fossil fuel reserves and the operation of semiconductor devices. Furthermore, very few undergraduate physics or chemistry texts discuss the quantum origins of the basis of bonding. Feynman, being a lover of fundamental physics, explains it very clearly in Chapter 10 of Volume III of his undergraduate lectures on physics [6].

To be strictly accurate, Feynman explains bonding through electron sharing by describing the hydrogen *ion*, which consists of two protons and one electron, because the mathematics is easier. But the uncertainty principle's responsibility for quantum bonding also works for two electrons and two protons as I describe. In both cases, an electron orbiting one proton moves more slowly when it can spread out over two protons, thanks to the uncertainty principle. Hence in the latter case the energy of motion of the electron and the total energy of the system are lower. That is the quantum bonding situation, which nature prefers.

Readers who would like to learn more about how quantum bonding explains the molecules of life and our fossil fuels, at a much simpler level than Feynman, will find the book in the 'Essential Chemistry' series by Phillip Manning helpful [7]. It has lots of pictures of the shapes electrons can take in their atomic orbits and only a few equations. The history of the development of quantum ideas to describe chemical bonds is covered in the intriguingly titled: *Neither Physics nor Chemistry* [8]. Undergraduates will find it stimulating to read the classic chemistry textbook by Linus Pauling [9], the double Nobel Prize-winner, who made such important contributions to explaining much of chemistry with quantum bonding ideas.

BIBLIOGRAPHY

1. Niels Bohr, *Essays 1932–1957 on Atomic Physics and Human Knowledge,* Ox Bow Press (1987).
2. Gavin Weightman, *Signor Marconi's Magic Box*, Harper Collins (2003).
3. http://www.nobelprize.org/nobel_prizes/physics/laureates/index. html, accessed 6 January 2014.
4. John Gribbin, *In Search of Schrödinger's Cat*, Black Swan (1991).
5. Hugh D. Young and Roger A. Freedman, *University Physics*, 11th edition, Pearson Education, Inc., Addison-Wesley (2004).
6. Richard P. Feynman, Robert B. Leighton and Matthew Sands, *The Feynman Lectures on Physics*, Volume III, Addison-Wesley (1965).
7. Phillip Manning, *Chemical Bonds*, Essential Chemistry, Chelsea House (2009).
8. Kostas Gavroglu and Ana Simoes, *Neither Physics nor Chemistry: a History of Quantum Chemistry*, Massachusetts Institute of Technology Press (2012).
9. Linus Pauling, *General Chemistry*, Dover (1988).

4. Brighter than a Thousand Suns

My history of the discovery of nuclear fission and the development of the first atomic weapons is based on Robert Jungk's fascinating book [1]. It was first published in German in 1956 and then in English in 1958. This is a definitive account because Jungk interviewed or corresponded with many of the scientists who were still alive. However, he was unable to interview physicists in the Soviet Union.

Jungk does not give much detail about the British contribution to the Manhattan Project or the sometimes fraught relationship between the US and the UK during the war. These are discussed in John Simpson's history of the post-war collaboration between Britain and the US on nuclear weapon development [2]. This book is also an important source of information on how 'Atoms for Peace' was a misnomer as far as the early UK civil nuclear programme was concerned.

The important part played by Frisch and Peierls in alerting Churchill to the danger that German scientists might develop a nuclear weapon first is described in their biographies in reference 3 of Chapter 2.

BIBLIOGRAPHY

More about the three types of radioactivity can be found in reference 1 of Chapter 1. The life stories of the physicists and chemists, who bravely researched the unknown properties of these radiations, can be found on the Nobel Foundation site, reference 3 in Chapter 3. Chadwick's account of the neutron discovery in his Nobel Prize lecture on this website is an interesting read. Chadwick does not admit to having the idea playing snooker. However, the mathematics he describes is equivalent to the equal-mass particle scattering problem. The latter is explained in many undergraduate physics textbooks, for example reference 5 of Chapter 3.

The data from France, Germany, Switzerland and the UK on the increased risk of childhood leukaemia within 5 km of a nuclear reactor has been reviewed by Alfred Koerblein and Ian Fairlie [3]. They show that the data on increased leukaemias from the four countries are similar and compatible. When they combine the data they find the increase is statistically significant, which indicates there is indeed an enhanced risk. The original authors of the German and Swiss studies concluded that the increase was statistically significant in their own data. In the case of the French and the UK studies, the original authors concluded that in their own data the increase was not statistically significant.

The secret early history of the UK civil nuclear and the link to the UK nuclear weapon programme was uncovered thanks to the persistence of a number of people, in particular the late Ross Hesketh [4]. Tony Benn was not told that plutonium from British civil reactors was being sent to the US under a military agreement while he was the Minister responsible for atomic energy [5].

The amount of weapon's grade plutonium which my colleagues and I calculated in 1985 (reference 7 in the Introduction) had been produced in British civil reactors in the early years of operation, was shown to agree with the government figure for plutonium from 'unidentified sites' in 2000 [6]. Reference 7 in that *Nature* correspondence article discusses the agreement between our 1985 figure for the UK civil plutonium sent to the US and the Clinton administration data. I have written a book chapter that gives further details of the civil–military plutonium link in the UK and more relevant references, including many from the parliamentary record [7].

BIBLIOGRAPHY

David Lowry and I examined the evidence as to whether history is repeating itself given that the Labour government's decisions on Trident renewal [8] and nuclear new build [9] were taken within five months of each other. We discussed the information that was publically available in an article in *New Statesman* in 2006 [10].

The July 2000 report detailing the large amounts of depleted uranium (more than 1,000 tons each year for the four years from 1979 to 1982) withdrawn from safeguards is not on the relevant government website [11]. This only carries information from 2001 onwards, though it does say the format is the same as used in the July 2000 report. The report is in the House of Commons Library. I received a paper copy from the House of Commons Information Office in 2012 [12].

The tragic legacy of cancers and birth defects in Iraq as a result of the use of depleted uranium in shells in the Gulf wars is described in the *Guardian* by the investigative journalist John Pilger [13]. The suggested mechanism, yet to be experimentally verified, which could make depleted uranium more dangerous than official risk assessments admit, is proposed in an article in the *Ecologist* by Chris Busby [14].

Information on the cost of the decommissioning and waste disposal for UK reactors, civil and military, can be found in references 30 and 31 of Chapter 10. Reference 32 in Chapter 10 charts the recent history of the large subsidy the government has devised to try to persuade investors to fund new reactors.

The Photovoltaic Power Systems Programme (PVPS) of the International Energy Agency [15] provides data on PV installations in different countries. The tabulated data which is cumulative to 2011 is presented in Table 2 of reference 15. The 2012 data is in Table 5 of reference 16.

1. Robert Jungk, *Brighter than a Thousand Suns: A Personal History of the Atomic Scientists*, Harcourt, Inc. (1958).
2. John Simpson, *The Independent Nuclear State: The United States, Britain and the Military Atom*, Macmillan (1983).
3. Alfred Koerblein and Ian Fairlie, 'French Geocap study confirms increased leukemia risks in young children near nuclear power plants', *International Journal of Cancer*, **131**, 2970 (2012).
4. Ross Hesketh, 'The export of civil plutonium', *Science and Public Policy*, **9**, 64 (1982).

BIBLIOGRAPHY

5. Tony Benn, *The End of an Era: Diaries 1980–90*, Arrow Books (1994).

6. Keith Barnham, Jenny Nelson and Rob Stevens, 'Did civil reactors supply plutonium for weapons?', *Nature*, **407**, 833 (2000).

7. Keith Barnham, Chapter 7 in *Plutonium and Security: The Military Aspects of the Plutonium Economy'*, ed. Frank Barnaby, Macmillan (1992), p. 110.

8. Ministry of Defence White Paper, 'The Future of the United Kingdom's Nuclear Deterrent', CM 6994, December 2006.

9. Department of Trade and Industry White Paper, 'Meeting the Energy Challenge: A White Paper on Energy', CM 7124, May 2007.

10. Keith Barnham and David Lowry, 'Strange Love', *New Statesman*, Energy Supplement, 15 May 2006.

11. Office for Nuclear Regulation, 'Withdrawal from Safe guards', http://www.hse.gov.uk/nuclear/safeguards/withdrawals.htm, accessed 6 January 2014.

12. 'Withdrawals from Safeguards pursuant to the UK Safeguards Agreement with the International Atomic Energy Agency (IAEA) and Euratom', DEP00/1261, referred to in Hansard, 28 July 2000, written answer, column 1094W. Available in 2012 from House of Commons Information Office, email: hcinfo@parliament.uk.

13. John Pilger, 'We've moved on from the war. Iraqis don't have that choice', *Guardian*, 27 May 2013.

14. Chris Busby, 'Uranium – the "demon metal" that threatens us all', *Ecologist*, 1 January 2014, http://www.theecologist.org/News/news_analysis/2205213/uranium_the_demon_metal_that_threatens_us_all.html, accessed 5 February 2014.

15. International Energy Agency, Photovoltaic Power Systems Programme (IEA-PVPS), 'Trends in Photovoltaic Applications', Report IEA-PVPS T1-21:2012, can be downloaded from http://www.iea-pvps.org/index.php?id=trends, accessed 5 February 2014.

16. International Energy Agency, Photovoltaic Power Systems Programme (IEA-PVPS), 'Trends 2013: in photovoltaic applications', Report IEA-PVPS T1-23:2013, can be downloaded from http://www.iea-pvps.org/index.php?id=trends, accessed 5 February 2014.

5. The Mystery of the Quantum Conductor

For a non-technical explanation of pre-semiconductor electronic devices, which operate using diode and triode valves inside vacuum tubes, I recommend *Electronics, The Life Story of a Technology* by David Morton and Joseph Gabriel [1]. The book also describes the development of the wide range of modern electronic devices made possible by semiconductors. It also contains some interesting personal recollections from the pioneers of the semiconductor revolution. Like many other books by technologists on the application and development of semiconductor devices, the revolutionary quantum ideas are not discussed. The reader is left with the impression that the transistor is simply a smaller and more robust version of the triode valve. This makes understanding the novel workings of modern devices like LEDs, solar cells and CMOS difficult.

Semi-Conductors and the Information Revolution: Magic Crystals That Made IT Happen by John Orton [2] is a book on the history of the semiconductor revolution that goes in more detail into the differences between metals, insulators and semiconductors. As a result, this book is clearer about how semiconductor devices work, as distinct from what they do. Like the Morton and Gabriel book it is intended for a non-scientific audience. John wrote what I consider to be the definitive history of the semiconductor revolution, which I relied on for Chapter 6. He describes this shorter book as the definitive history 'with all the mathematics and much of the physics removed'.

As I mention in Chapter 6 and in the Acknowledgements, I owe John Orton a debt of gratitude for allowing me to join his group at Philips Research Laboratories (PRL) in Redhill. He and the members of his group were most helpful in explaining state-of-the-art semiconductor structures. I enjoyed collaborating with a number of them in subsequent years.

Books that explain the differences between conductors, insulators and semiconductors at the quantum level in general use mathematics. My non-mathematical descriptions follow the explanations in Feynman's undergraduate lectures, which are in reference 6 of Chapter 3. This is where the epigraph at the start of the chapter originated.

If you read any of these books you will need to bear in mind that some of the names I use are unique to this book and to Silicon Street. To avoid confusion here is a handy glossary:

> *Dual-conductor* = semiconductor
> *Positron* = hole
> *Diversity atom* = impurity atom = dopant
> *Diversity diode* = semiconductor diode = p-n junction

1. David Morton and Joseph Gabriel, *Electronics: The Life Story of a Technology*, Johns Hopkins University Press (2007).
2. John Orton, *Semiconductors and the Information Revolution: Magic Crystals That Made IT Happen*, Academic Press (2009).

6. The Semiconductor Revolution

A fascinating account of the discovery of the transistor has been written by a co-inventor, Walter Brattain. It contains copies of laboratory notes and original papers, notes for non-specialists and an autobiography [1]. I took the quotation at the start of the chapter from the autobiography. Sadly it does not appear that this journal is available on the web. A similar autobiography, with an almost identical quotation, appears in reference 2.

A news story from 2010 about the first two silicon chips with a billion transistors can be found in reference 3.

I recommended John Orton's book on the history of semiconductors for non-scientists in the last chapter. This chapter leans very heavily on John's definitive *History of Semiconductors* [4]. It goes in detail into the developments of the devices of the semiconductor revolution, such as integrated circuits, CMOS memory, LEDs, lasers and the quantum wells that led to the mobile phone.

For a more technical description of the physics of solar cells themselves, I can thoroughly recommend the textbook written by Jenny Nelson [5]. Jenny, as the Greenpeace Fellow, co-founded the Quantum Photovoltaic Group at Imperial with me in 1989.

The history of solar cells from the pioneering work through to the take-off in Germany and Japan at the end of the twentieth century is described for non-specialist readers by Perlin [6].

BIBLIOGRAPHY

The performance of the triple-junction concentrator cell with QuantaSol's technology in the middle cell is described in reference 8 of Chapter 9.

1. Walter Brattain, 'Discovery of the Transistor Effect: One Researcher's Personal Account', *Adventures in Experimental Physics*, **5**, 1, (1976).
2. See also Walter Brattain, 'Offspring of the physics revolution', *Electronics and Power*, **19**, 58 (1973). http://ieeexplore.ieee.org/xpl/login.jsp?tp=&arnumber=5180748&url=http%3A%2F%2Fieeexplore.ieee.org%2Fiel5%2F5176125%2F5180709%2F05180748.pdf%3Farnumber%3D5180748, accessed 5 February 2014.
3. John Stokes, Arstechnica, 'Two billion-transistor beasts: POWER7 and Niagara 3', 10 February 2010, http://arstechnica.com/business/2010/02/two-billion-transistor-beasts-power7-and-niagara-3/, accessed 5 February 2014.
4. John Orton, *The Story of Semiconductors*, Oxford University Press (2004).
5. Jenny Nelson, *The Physics of Solar Cells*, Imperial College Press (2003).
6. John Perlin, *From Space to Earth: The Story of Solar Electricity*, Harvard University Press (2002).

7. Questions for the Solar Revolution

I have described a very simplified picture of the physics of wind. My main aims were to show how complementary wind power is to photovoltaic power and to be able to explain how global warming is leading to stronger storms. Those who wish to learn more about the global wind systems should consult the *Philip's Guide to Weather* written for a popular audience with many helpful diagrams [1]. For more detail about how temperature differences drive winds and the physics of the atmosphere and climate change I recommend *Atmosphere, Clouds and Climate* [2].

Those who would like to learn more about both the history of wind power and modern developments, should read Peter Musgrove's book

BIBLIOGRAPHY

[3]. The book has a very useful global overview of the take-up of wind power.

You can find more detailed information supporting my answers to the questions in a paper that I wrote with Kaspar Knorr of the Fraunhofer Institute for Wind Energy and Energy System Technology, Kassel, Germany and Massimo Mazzer of the CNR-IMEM in Parma, Italy [4]. The paper was published in the journal *Energy Policy*. In Figure 3 the actual power demand of the German grid is shown scaled-down for a typical week in September 2006. The electrical power output from the PV, wind and biogas generators of the Kombikraftwerk experiment that matched the demand are also shown. The data came from Kaspar who is in charge of the ongoing Kombikraftwerk experiment. There is even more detailed information on the first very successful year of the experiment in reference 5.

Figure 2 in the *Energy Policy* paper shows how the peak wholesale price of electricity on the German grid has declined compared to the night-time price as the amount of PV power on the grid has increased [6]. The paper also has graphs showing that peak daytime electricity demand is near noon in both Britain and Germany, summer and winter. So a similar fall in the peak daytime price of electricity should occur also in the UK when Britain has as much PV as Germany.

The observation that, on average, sunlight in the UK contains only 5 per cent less energy than in Germany is taken from data in a publication by the team that developed the Photovoltaic Geographical Information System (PVGIS) [7].

I first heard about the Kombikraftwerk experiment on reading *The Solar Century*, by Jeremy Leggett [8]. This book is an excellent source of information on all practical aspects of photovoltaics. It has some great pictures of imaginative applications and suggestions of how the solar revolution could develop via local grid networks. As well as being complimentary to *The Solar Century*, I have tried to keep *The Burning Answer* complementary in its coverage of PV.

Reference 9 reports that the British National Grid will have to spend over one billion pounds to strengthen the UK electricity grid, if new reactors are built at Hinkley Point.

Kaspar Knorr, Massimo Mazzer and I wrote a correspondence piece in *Nature Materials* [10] summarising our *Energy Policy* paper. Looking

BIBLIOGRAPHY

at the performance of PV on the German, Italian and British electricity grids we concluded that a moratorium on all new electricity generation apart from the renewables was now possible. It would be the most effective way to reduce carbon levels and also lead to cheaper electricity.

The exponential expansion of the PV installations in Germany can be seen in figures in both the *Nature Materials* [10] and *Energy Policy* [4] papers. The data was taken from reference 15 of Chapter 4. The average year-on-year expansion of PV installations in Germany was 55 per cent for the five years to 2010. The average year-on-year expansion of mobile phone subscribers in OECD countries was 48 per cent for the five years to 2000 [11].

Details of the German feed-in-tariff (FIT), which produced this exponential increase in PV and has supported expansion of other renewables in Germany, can be found in a recent paper by Amory Lovins [12].

The predictions about the future expansion of the PV and wind supplies in Germany and the UK are discussed in the *Energy Policy* article. Massimo Mazzer and I made the observation that Britain was around nine years behind Germany in onshore wind power installations in 2006 [13]. In that paper we predicted that if the UK followed the German policy, so that wind power in Britain expanded at the rate Germany had achieved nine years earlier, then by 2020 the UK would have 25 GW of onshore wind power installed. Sadly such policies have not been implemented. By the end of 2012, the UK only had just under 6 GW of onshore wind power installed [14]. Germany achieved this level in 2000 [15, 16].

In fact, Germany took only nine years to move from the 6 GW level to install over 25 GW of onshore wind in 2009 [15, 16]. So I can update our earlier prediction and say that were the German performance replicated in the UK from 2014 onwards, the 25 GW level of onshore wind power could be achieved by 2023.

A way to counter the power of the anti-wind lobby, which has been successfully employed in Germany, is discussed in Chapter 9 reference 28. The UK government prediction of 33 GW of offshore wind by 2020 was reported in reference 17.

1. Ross Reynolds, *Philip's Guide to Weather*, Octopus Publishing Group (2006).

BIBLIOGRAPHY

2. David Randall, *Atmosphere, Clouds and Climate*, Princeton University Press (2012).

3. Peter Musgrove, *Wind Power*, Cambridge University Press (2010).

4. Keith Barnham, Kaspar Knorr and Massimo Mazzer, 'Benefits of photovoltaic power in supplying national electricity demand', *Energy Policy*, **54**, 385 (2013).

5. Kombikraftwerk, 'Technical Summary of the Combined Power Plant' (2007), http://www.kombikraftwerk.de/fileadmin/downloads/Technik_Kombikraftwerk_EN.pdf, accessed 6 January 2014.

6. Christoph Podewils, Philippe Welter and Chris Warren, 'Downward Trend', *Photon International*, August 2011, 64. http://www.photon-magazine.com/.

7. M. Suri, T.A. Huld, E.D. Dunlop and H.A. Ossenbrink, 'Potential of solar electricity generation in the European Union member states and candidate countries', *Solar Energy*, **81**, 1295 (2007). See also http://re.jrc.ec.europa.eu/pvgis.

8. Jeremy Leggett, *The Solar Century: The past, present and world-changing future of solar energy*, GreenProfile (2009).

9. Alex Froley, 'UK's National Grid to spend $1.6 billion connecting Hinkley Point Nuclear Plant', Platts, 21 November 2013, http://www.platts.com/latest-news/electric-power/london/uks-national-grid-to-spend-16-billion-connecting-26481166, accessed 5 February 2014.

10. Keith Barnham, Kaspar Knorr and Massimo Mazzer, 'Progress Towards an All-Renewable Electricity Supply', *Nature Materials*, **11**, 908 (2012).

11. OECD Key ICT Indicators 2013, 'Mobile subscriptions in total for OECD (millions)', http://www.oecd.org/internet/broadband/oecdkeyictindicators.htm, accessed 7 January 2014.

12. Amory Lovins, 'Renewable Germany: The Very Model of a New Energy Order', 18 April 2013, http://reneweconomy.com.au/2013/renewable-germany-the-very-model-of-a-new-energy-order-53357, accessed 7 January 2014.

13. Keith Barnham and Massimo Mazzer, 'Vorsprung durch technik', *New Statesman*, Energy Supplement, 15 May 2006.

14. UK Government, Digest of UK energy statistics (DUKES), Chapter 6, 'Renewable sources of energy', https://www.gov.uk/government/

uploads/system/uploads/attachment_data/file/65850/DUKES_2013_Chapter_6.pdf, accessed 19 January 2014.

15. Volker-Quaschning, 'Installed capacity of renewable power plants in Germany in MW', http://www.volker-quaschning.de/datserv/ren-Leistung-D/index_e.php, accessed 7 January 2014.

16. German Wind Energy Institute, *DEWI Magazin*, no. 38, p. 38, February 2011, http://www.dewi.de/.

17. Energy Efficiency News, 'UK government confirms extra 25 GW of offshore wind energy', 25 June 2009, http://www.energyefficiencynews.com/articles/i/2205/, accessed 8 January 2014.

8. The Solar Cornucopia

I found the Renewable Energy Association's report 'Renewable Energy: Made in Britain' [1], extremely helpful and can thoroughly recommend it as a starting point for practical information on all the solar technologies. If you don't find a reference for a particular point in this and the next chapter, it probably came from reference 1.

I deliberately misquoted the title of Al Gore's book [2]. This was of course a definitive work in alerting the world of the need for a solar revolution. His title referred to the problem. It is extremely convenient that we have such a range of renewable technologies to provide many solutions. Gore's book was particularly important in exposing the efforts of the fossil fuel industry to deny global warming. My intention is that *The Burning Answer* provides the information to counter the arguments of those who deny that the solar technologies can do the job.

The principle that 'small is beautiful' was the title of the groundbreaking book on people-scale economics by E.F. Schumacher [3]. First published in 1973, it contained a chapter challenging the expansion of nuclear power, mainly on the grounds of the (still unsolved) waste problem.

The UK Department of Energy and Climate Change (DECC) roadmap in 2011 did not include either geothermal or photovoltaics in its top eight technologies [4]. Professor David MacKay's book is reference 5. His analogy between geothermal heat extraction and sucking a drink out of crushed ice appears on page 96. On page 97 he says of the earth's natural geothermal energy that 'the heat flow at the surface is 50 mW/

m²'. On the next page he summarises his calculation of the 'sustainable forever' heat flow extracted from the optimal depth. He says 'an ideal heat engine would deliver 17 mW/m²'. This is less than the natural geothermal heat flow at the surface. Ladislaus Rybach wrote the paper with a full calculation of sustainable geothermal heat extraction allowing for the heat that flows back [6].

The consultants Sinclair, Knight and Merz produced the report that calculated that geothermal energy could produce 20% of the UK electrical power and supply all the domestic heating [7]. This suggests that schemes like the ones that are supplying supermarkets with all their heating and cooling (Chapter 9 reference 23) could clearly be extended to industrial and public buildings throughout the UK without worrying about sustainability.

James Lovelock's book, *The Revenge of Gaia,* is reference 8.

The sad tale of Salter's Duck is an example of the negative influence of the nuclear and fossil fuel lobbies on renewable energy developments in the UK. The history is described by David Ross in a paper in Science and Public Policy [9].

Information about the current UK companies that are leading the way in wave and tidal power was again from the Renewable Energy Association [1]. My old friend John Hassard provided the update on the Hydro-Venturi technology [10].

The negative environmental impacts of using biofuels from biomass for transportation have been discussed by George Monbiot [11]. The Renewable Energy Association review [1] paints a more positive picture of recent developments of these technologies. The German Federal Environment Agency regards the use of biomass for electricity generation (or heat generation) as unnecessary. Their opinion is that sufficient can be produced from organic wastes and residues [12].

The Energy Savings Trust can be contacted at reference 13. The report that the number of domestic loft insulation refurbishments had fallen by 93 per cent in 2013 appeared in the *Guardian* [14].

The Committee on Climate Change proposed a carbon limit for electricity generation in 2030 of 50 gCO_2/kWh [15]. The coalition government does not want to set any limit. Were it to do so, it would prefer 450 gCO_2/kWh, which is nine times higher [16]. This is to allow natural gas electricity generators to be built.

BIBLIOGRAPHY

Daniel Nugent and Benjamin Sovacool have reviewed published LCAs for wind and PV energy generation [17]. The three reviews of the published life cycle analyses for nuclear power are by Benjamin Sovacool [18], Ethan Warner and Garvin Heath [19] and Jef Beerten and colleagues [20]. The report reviewing LCAs for the Committee on Climate Change prepared by Ricardo-AEA can be found at reference 21.

The cost of construction of a hydropower dam was taken from reference 3 of Chapter 9 and of the EPR from reference 4 in Chapter 16.

Jan Willem Storm van Leeuwen explained to me the importance of the very strong dependence of the lifetime carbon emissions of nuclear power on the concentration of uranium oxide in the ore. His reports can be found at reference 22.

1. Renewable Energy Association, 'Renewable Energy: Made in Britain' (2012), http://www.r-e-a.net/.
2. Al Gore, *An Inconvenient Truth*, Bloomsbury (2006).
3. E.F. Schumacher, *Small Is Beautiful: A Study of Economics As If People Mattered*, Sphere Books (1974).
4. UK Department of Energy and Climate Change (DECC), 'UK Renewable Energy Roadmap', 2011, https://www.gov.uk/government/uploads/system/uploads/attachment_data/file/48128/2167-uk-renewable-energy-roadmap.pdf.
5. David MacKay, *Sustainable Energy – Without the Hot Air*, UIT Cambridge Ltd (2009).
6. Ladislaus Rybach, 'Geothermal Sustainability', *GHC Bulletin*, September 2007, Geo-Heat Center, Exploration/Evaluation, http://geoheat.oit.edu/ghcindex.htm.
7. Sinclair, Knight and Merz, 'Geothermal Energy Potential in Great Britain & Northern Ireland', 26 May 2012, http://www.globalskm.com/Insights/News/2012/SKM-report-on-Geothermal-Energy-Potential-in-Great-Britain--Northern-Ireland.aspx, accessed 5 February 2014.
8. James Lovelock, *The Revenge of Gaia*, Penguin Books (2007).
9. David Ross, 'Scuppering the waves: how they tried to repel clean energy', *Science and Public Policy*, **29**, 25 (2002).
10. John Hassard, Imperial College London, private communication. See also http://www.hydroventuri.com/.

BIBLIOGRAPHY

11. George Monbiot, 'Must the poor go hungry just so the rich can drive?' *Guardian*, 14 August 2012, http://www.monbiot. com/2012/08/13/hunger-games/, accessed 8 January 2014.

12. German Federal Environment Agency, 'Energy target 2050: 100% renewable electricity supply', Umwelt Bundes Amt (UBA) (July 2010), http://www.umweltdaten.de/en/publikationen/weitere_infos/3997-0.pdf.

13. Energy Savings Trust, http://www.energysavingtrust.org.uk/, accessed 5 February 2014.

14. Damian Carrington, 'Number of Homes Taking Energy Saving Measures Falls', *Guardian*, 31 December 2013.

15. UK Committee on Climate Change, 'Meeting Carbon Budgets – 2012 Progress Report to Parliament' (June 2012), Key Findings, http://www.theccc.org.uk/publication/meeting-the-carbon-budgets-2012-progress-report-to-parliament/.

16. UK government press release, 'Davey sets out measures to provide certainty to gas investors', 17 March 2012, https://www.gov.uk/government/news/davey-sets-out-measures-to-provide-certainty-to-gas-investors, accessed 12 January 2014.

17. Daniel Nugent and Benjamin Sovacool, 'Assessing the lifecycle greenhouse gas emissions from solar PV and wind energy: A critical meta-survey', *Energy Policy*, 65, 229 (2014).

18. Benjamin Sovacool, 'Valuing the greenhouse gas emissions from nuclear power: A critical survey', *Energy Policy*, 36, 2940 (2008).

19. Ethan Warner and Garvin Heath, 'Life Cycle Greenhouse Gas Emissions of Nuclear Electricity Generation: Systematic Review and Harmonization', *Journal of Industrial Ecology*, 16, S73 (2012).

20. Jef Beerten, Erik Laes, Gaston Meskens and William D'haeseleer, 'Greenhouse gas emissions in the nuclear life cycle: A balanced appraisal', *Energy Policy*, 37, 5056 (2009).

21. Naser Odeh, Nikolas Hill and Daniel Forster, 'Current and Future Lifecycle Emissions of Key "Low Carbon" Technologies and Alternatives', Ref. ED58386, 17 April 2013, Ricardo-AEA Ltd, Harwell, Didcot, OX11 0QR, http://www.theccc.org.uk/wp-content/uploads/2013/04/Ricardo-AEA-lifecycle-emissions-low-carbon-technologies-April-2013.pdf, accessed 20 January 2014.

BIBLIOGRAPHY

22. Jan Willem Storm van Leeuwen, 'Nuclear Power, Energy Security and CO2 Emission', May 2012 , http://www.stormsmith.nl/reports.html, accessed 20 January 2014.

9. How Can We Reduce Our Carbon Emissions?

January 2009 was a tipping point for the cost of solar panels and for the solar revolution. This can most clearly be seen on the Solarbuzz site [1] in a diagram which had two lines tracking the retail price of a PV panel in Europe and the US. Sadly Solarbuzz stopped updating this figure in early 2012. The whole of the fall to below $1/W in June 2012 can be seen in Figure 9 of the Lawrence Berkeley National Laboratory publication 'Tracking the Sun VI' [2]. This figure also shows that, in the USA, the balance of system costs had not fallen as fast. Hopefully, this will fall faster in the near future as there are signs PV sales are picking up at last in the US.

I have taken construction costs for all electrical generation technologies, apart from the rapidly falling solar cost, from the Mott MacDonald report, which was prepared for the UK Department of Energy and Climate Change (DECC) in 2010 [3]. The onshore wind construction cost in Figure 6.2 of this report was just over $2/W. There is more up-to-date data on the DECC website, but it is not easy to extract capital costs from them. Most of the construction costs for electrical generation have not changed much from 2010 with two notable exceptions. First, thanks to the formation of a PV market and the exponential growth in demand, the cost of new PV plant has declined fast as shown in references 1 and 2. Second, the construction cost of the French prototype of the EPR reactor rose from $2/W in 2005 to $7/W in December 2012 [4]. We will find in reference 4 in Chapter 16 that a year later, in October 2013, the 1.6 GW EPR that EDF intend to build at Hinkley Point C was expected to cost £8 billion. Assuming $1.5/£ the next EPR will cost $7.5/W. I am sure this will rise even further when one or other of the two prototypes work. Nuclear reactors are too large for mass production or for a market to form, so prices will always tend to rise.

I used the comparemysolar website [5] to select three companies, which I asked to visit my home and provide written quotations for PV

344

on the roof. I recommend you check if a company provides its own guarantee that they will pay the scaffolding cost in the unlikely event that a panel develops a fault and needs replacing during the manufacturer's warranty period.

Your PV supplier should be able to install one of the boxes of electronics that heat water with the excess electricity from your PV. I am aware of two companies manufacturing them, ImmerSun [6] and SOLiC [7]. Make sure the box will trickle the excess to the immersion. You don't want this backed up to full power from the grid.

A picture of a computer simulation of the multifunctional smart windows system, which generates electricity from the direct sunlight but allows diffuse light into the room, is shown in a figure in a paper I wrote with colleagues for the trade magazine *Compound Semiconductor* [8].

I used the Green Electricity Market Place [9] to decide Good Energy were the all-renewable electricity supplier for me. I am grateful to Ed Gill for explaining the company philosophy. Further details can be found on their website [10]. Ed is no longer with the company, but Chris Welby kindly confirmed some points [11].

Information on solar thermal systems can be found from the Energy Savings Trust website [12]. The Trust's statement about heating with electricity, which I argue is now out of date, is taken from reference [13] where you can also find information on modern efficient electric storage heaters. Information on the comparative efficiencies of gas boilers can be found on the SAP900 'Boiler Efficiency Database' [14].

A neighbour, John Warman, has been pleased with the electrical central heating system he purchased from Ecopowerheating [15]. He had also researched and recommends Ecowarmth [16]. Both websites have online quotation facilities.

Reference 1 of Chapter 2 describes the life and achievements of Lord Kelvin, the inventor of the heat pump. One of the most interesting parts of the book is the way his clearly very creative understanding of the nature of heat led him to challenge Maxwell's approach to the electromagnetic field. I see this as a forerunner to the controversies over quantum ideas early in the next century.

A simple description of how the ground source and air source heat pumps work can be found on the website of one manufacturer, Dimplex [17]. There is a good animation on YouTube [18]. I found the information that as early as 2002 over 90 per cent of the new build homes in Sweden had heat pumps in reference 19.

Information on costs of ground source and air source heat pumps can be found from the *Which?* comparison sites [20, 21]. The air source site has a picture of a system, which you can see closely resembles an air-conditioning unit. The Energy Savings Trust website, which was reference 13 in Chapter 8, also has much information on heat pumps. I have found their enquiry service [22] extremely helpful. The website for the company Greenfield Energy, which is using oil-industry approaches to extract geothermal energy under supermarket car parks, is at reference 23.

Much of the information on the greening of the UK gas grid came from the REA report, which is reference 1 in Chapter 8, and from Ciaran Burns at the REAL Green Gas Certification scheme [24]. The information on the first commercial biomethane to gas-grid plant was provided by Nick Finding the CEO of JVEnergen [25]. One example of an alternative approach to waste to gas generation by gasification is described on the Advanced Plasma Power website [26]. Encouraging predictions for the potential of anaerobic digestion, biodegradable waste and other forms of waste to replace natural gas on the UK domestic gas grid can be found in the 2009 report from the National Grid [27].

Renewable Energy Focus carried the story about the remarkable percentages of co-operatively owned wind turbines in Denmark and Germany [28]. The article also discusses Feldheim in Germany, which has 150 inhabitants and which claims to be the world's first carbon neutral village. Details are also provided of the community owned wind turbine in the Forest of Dean in the UK, which was approved, while objections were upheld for a nearby turbine, which was to have been installed by a commercial company [28].

A number of examples of the over 100 community-led renewable energy schemes that are supported by the business-led charity Pure Leapfrog appear on their website [29]. They include the Cheshire village Ashton Hayes, which aims to be the first carbon neutral village in the UK [30]. They also support the Repowering London [31] project in a deprived, inner city area of Brixton in south London. Another different example is provided by the Bath and West Community Energy

BIBLIOGRAPHY

community benefit society, which has raised over a million pounds through two share issues to local residents [32].

1. Solarbuzz, 'Module prices', still accessible 10 January 2014 though not updated, http://www.solarbuzz.com/facts-and-figures/retail-price-environment/module-prices.
2. Galen Barbose, Naïm Darghouth, Samantha Weaver and Ryan Wiser, 'Tracking the Sun VI: An Historical Summary of the Installed Price of Photovoltaics in the United States from 1998 to 2012', Lawrence Berkeley National Laboratory, July 2013, http://emp.lbl.gov/sites/all/files/lbnl-6350e.pdf, accessed 10 January 2014.
3. Mott MacDonald, 'UK Electricity Generation Costs Update', June 2010, https://www.gov.uk/government/uploads/system/uploads/attachment_data/file/65716/71-uk-electricity-generation-costs-update-.pdf, accessed 10 January 2014.
4. Muriel Boselli and Michel Rose, 'EDF raises French EPR reactor cost to over \$11 billion', Reuters, 3 December 2012, http://www.reuters.com/article/2012/12/03/us-edf-nuclear-flamanville-idUS-BRE8B214620121203, accessed 12 January 2014.
5. Comparemysolar PV price comparison site, http://comparemysolar.co.uk/ accessed 5 February 2014.
6. Immersun, 'Next generation Immersun' http://www.immersun.co.uk/, accessed 16 January 2014.
7. The SOLiC system can ordered from Earthwise Products Ltd, http://www.earthwiseproducts.co.uk/info/solic-200+73.html, accessed 16 January 2014.
8. Keith Barnham, Alison Dobbin, Matt Lumb and Tom Tibbits, 'Tuning the Triple-junction', *Compound Semiconductor*, March 2011, p. 32, http://www.compoundsemiconductor.net/csc/features-details.php?cat=features&id=19733453&key=Tuning%20the%20triple-junction&type, accessed 20 January 2014.
9. The Green Electricity Market Place is at http://www.greenelectricity.org/index.php, accessed 16 January 2014.
10. Good Energy website, http://www.goodenergy.co.uk/about/a-different-kind-of-energy-company/how-we-buy-and-sell-electricity, accessed 16 January 2014.

BIBLIOGRAPHY

11. Chris Welby, Policy and Regulatory Affairs Director, Good Energy, private communication.
12. Energy Savings Trust, 'Solar Water Heating', http://www.energysavingtrust.org.uk/Generating-energy/Choosing-a-renewable-technology, accessed 17 January 2014.
13. Energy Savings Trust, 'Improving Electric Systems', http://www.energysavingtrust.org.uk/Heating-and-hot-water/Improving-electric-systems, accessed 17 January 2014.
14. Standard Assessment Procedures for the Energy Ratings of Dwellings, SAP900, 'Boiler Efficiency Database', http://www.boilers.org.uk/pages/hw.htm, accessed 17 January 2014.
15. Ecopowerheating, 'Performance, Efficiency, Design', http://www.ecopowerheating.co.uk/, accessed 17 January 2014.
16. Ecowarmth, 'The Specialists in German Electrical Heaters', http://www.ecowarmth-ltd.com/, accessed 17 January 2014.
17. See the helpful diagram at http://www.dimplex.co.uk/products/renewable_solutions/heat_pumps_explained.htm, accessed on 18 January 2014.
18. Chaffeeair, 'How does a heat pump work?', see video at http://www.youtube.com/watch?v=g39nM7GbSJA.
19. 'World Energy News', *Heat Pump News*, Issue 1, Autumn 2002, http://www.heatpumps.org.uk/PdfFiles/HeatPumpNewsNo.1.pdf, accessed 18 January 2014.
20. Information on ground source heat pumps: http://www.which.co.uk/energy/creating-an-energy-saving-home/guides/ground-source-heat-pumps-explained/, accessed 18 January 2014.
21. Information on air source heat pumps: http://www.which.co.uk/energy/creating-an-energy-saving-home/guides/air-source-heat-pumps-explained/, accessed 18 January 2014.
22. Information on heat pumps and comparison with other heating approaches can be requested from energy-advice@est.org.uk.
23. Greenfield Energy, 'Sustainable Thermal Energy Networking', http://www.geoscart.com/.
24. Renewable Energy Association Limited, 'Green Gas Certification Scheme', http://www.greengas.org.uk/, accessed 18 January 2014.
25. JVEnergen, 'Welcome to Rainborrow Farm Anaerobic Digester Plant', http://www.jvenergen.co.uk/, accessed 18 January 2014.

26. Advanced Plasma Power, 'Transforming Waste into Energy and Fuels', http://www.advancedplasmapower.com/, accessed 18 January 2014.

27. National Grid, 'The Potential for Renewable Gas in the UK', January 2009, http://www.nationalgrid.com/NR/rdonlyres/9122AEBA-5E50-43CA-81E5-8FD98C2CA4EC/32182/renewablegasWPfinal1.pdf, accessed 6 February 2014.

28. George Marsh and Liz Nickels, 'Community, Crowd and Conversion', Renewable Energy Focus, 29 August 2013, http://www.renewableenergyfocus.com/view/34243/community-crowd-and-conversion/, accessed 18 January 2014.

29. Pure Leapfrog, 'Energising Communities', http://www.pureleapfrog.org/, accessed 18 January 2014.

30. Ashton Hayes Community Energy, 'Ashton Hayes Going Carbon Neutral', http://www.goingcarbonneutral.co.uk/ashton-hayes-community-energy/, accessed 18 January 2014.

31. Repowering London, 'Creating Local Energy', http://www.repowering.org.uk/, accessed 18 January 2014.

32. Bath and West Community Energy, 'Generating Local Energy', http://www.bwce.coop/, accessed 18 January 2014.

10. The Politics of the Solar Revolution

Amory Lovins describes recent progress in the German renewable energy industry including its contribution to the speed at which that country is recovering from the 2008 crash in reference 12 of Chapter 7. The German government's pathway towards an all-renewable electricity supply by 2050 is described in reference 12 of Chapter 8. This policy is often described as an 'Energiewende' or 'energy turn', but as one account of the history [1] relates, the name has a 30-year history.

Renewables already supply 99 per cent of Norway's electricity production [2] and 99.9 per cent of Iceland's electricity production [3]. Denmark plans to generate all their energy renewably and be independent of fossil fuels by 2050 [4]. Their pioneering community owned heat and power systems are described in reference [5]. Scotland's target is 100 per cent of electricity demand being supplied by renewables in 2020 [6].

BIBLIOGRAPHY

David MacKay claimed that 'a third of the land' would need to be covered by solar cells in an interview with the Environment Editor of *The Sunday Times* [7]. Professor MacKay's book is reference 5 in Chapter 8. The picture of a Cambridgeshire rooftop delivering 20 W per square metre of panel is on page 40. The solar PV farm delivering 5 W/m^2 is on page 41. Here, under the heading 'Fantasy time: solar farming', you can also find two other numbers that were important in my reconciliation of Professor MacKay's figure for the area of the UK required by PV with my own. He says that a solar cell of 10% efficiency would deliver power at 10 W/m^2. Note this is an average power. This is one tenth of the peak power. On the same page he quotes the total UK land area as 4000 m^2 per person and the land area of buildings as 48 m^2 per person. The latter is 1.2 per cent of the former.

The references for my claim that PV on south-facing roofs could supply all the electrical energy consumed inside the buildings are cited in my *Nature Materials* commentary article, which is reference 3 in Chapter 15. George Monbiot made his criticism of Jeremy Leggett in his book *Heat: How to Stop the Planet Burning* [8].

Monbiot strongly opposed the first version of the UK feed-in-tariff a month before it was introduced in 2010 [9]. The FIT was an immediate success. It was severely cut in 2011, but by end 2012 nearly 2 GW of PV power had been installed on the UK grid. This is more power than the nuclear reactor EDF hopes to have operational in 2023.

In 2009, Professor David MacKay predicted the UK would have 0.75 GW of PV in 2050 if it expanded at 'the maximum rate I think is plausibly achievable'. This is one of many predictions made in a memorandum to the House of Commons, Environmental Audit Committee in 2009 [10]. In fact MacKay's figure of 0.75 GW in the memorandum is an average power. To get the installed PV power in 2050, which is what matters in satisfying demand, I have multiplied MacKay's estimate by ten. That is because the sun only shines about 10 per cent of the year on average in the UK. According to MacKay, the most PV that Britain could hope for in 2050 is an installed power of 7.5 GW. Assuming PV in the UK expands at the rate achieved in reaching Germany's 7.5 GW from the same level as Britain had in 2008, this target could be achieved by 2019 – 31 years earlier than is possible at MacKay's 'plausibly achievable' maximum rate. Reference 15 of Chapter 4 has the installation data for Germany and the UK.

BIBLIOGRAPHY

The computer program 'Pathways to 2050' produced by the UK Department of Energy and Climate Change (DECC) is described in the report '2050 Pathways Analysis' from which the quotation about matching *energy* supply and demand was taken [11]. The implications of the basis being energy flows rather than power, which is normally used for supply and demand, are not discussed in the report. Further evidence that the methodology is based on energy flows comes from the Pugwash report [12]. They describe how to obtain the 'Sankey' plots of their pathways from the DECC spreadsheets. Sankey plots are energy flow diagrams. The information on the 'stress test', the only use of power in the programme, I also got from the Pugwash report.

You can learn about the zero carbon nuclear assumption and the time for which nuclear reactors are assumed to run each year, by downloading the examples of DECC's Pathways EXCEL spreadsheets in the Pugwash report [12] and clicking on the tag 'constants'. The information that in practice between 2008 and 2012 British nuclear reactors only delivered power 50–70 per cent of the time can be found from reference 13.

Details of the budgets of the US national laboratories can be downloaded from the web [14]. The privatisation of the PV section of the UK National Renewable Energy Centre (Narec) and its 15 employees is described in a Narec press release [15]. Information on the Fraunhofer ISE and IWES laboratories is at reference 16. The UK National Nuclear Laboratory has 550 research staff according to Chapter 2 of a recent House of Lords report [17].

Quantasol's briefly held world record for a single junction concentrator cell of 28.3 per cent at 534 times solar concentration is described in reference 18.

Lord May criticised the climate change deniers in the *Guardian* [19]. He also criticised NGOs opposed to nuclear power in his final speech as president of the Royal Society [20]. His opinion that wind and solar supplying our energy needs was 'wishful thinking' appeared in the *Daily Telegraph* [21].

The two Royal Society reports that I criticise can be downloaded from the society website [22, 23]. The report on the alternative immobilisation approach for plutonium can be found in an article in *Nature* [24], which also discusses the problems with mixed-oxide fuel production in the UK and the US. The opinion of the *Independent* on hearing of the

BIBLIOGRAPHY

less than 2 per cent throughput in the first five years' operation of the plant is at reference 25. The wise words on plutonium from the Royal Commission on Environmental Pollution can be found in reference 3 of Chapter 1.

The joint report by the Royal Society and the Royal Academy of Engineering on shale gas extraction can be downloaded from the Royal Society website [26]. The International Energy Agency view that only one-third of proven fossil fuel reserves can be burnt before 2050 if we are to avoid the worst dangers of global warming can be found in reference 27. The University of Cornell paper on methane leaks is reference 28. Evidence of radiation above recommended levels in wastewater has been reported as a result of fracking in Pennsylvania [29].

The UK government's commitment to nuclear waste disposal has been running at around £2 billion a year for some time. This has recently been upped to £3 billion a year [30]. The total commitment is unclear. This is not surprising considering the uncertainty over how the waste is finally to be stored. Estimates vary between £58.9 billion and £104 billion for the total commitment, depending on the accounting method. This can be seen on page 20 of the Nuclear Decommissioning Authority's accounts [31]. Another concern from this page is that there does not appear to be any provision for the decommissioning of the second-generation AGR reactors.

I have found the blog maintained by David Tokes [32] helpful in keeping up to date on the protracted wrangling between the British government and EDF (in which the French government have a very large shareholding) over the 'contracts for difference'. This is effectively a subsidy which taxpayers will have to fund. It is possible that it will violate EU competition rules.

The BP group does not keep press releases from before 2000 so I have been unable to provide a verifiable reference for John Browne's announcement. First Solar was the first PV company to sell 1 GW of PV panels in one year, having broken through the $1/W cost barrier earlier in 2009 [33]. BP announced in 2011 that they were closing down all their solar panel production, while continuing to spend $20 billion a year on oil and gas development [34]. The net sales for First Solar in 2011 were $2.8 billion [35] while BP's turnover in 2011 was $376 billion [36].

BIBLIOGRAPHY

According to the Renewable Energy Association (reference 1 in Chapter 8), the US fossil fuel industry spends 20 times as much on lobbying the government as does the renewable energy industry. Given the relative turnovers of the leading, non-Chinese, PV company, First Solar ($2.8 billion) and BP ($376 billion), I would not be surprised if the ratio was much higher.

The news that David Cameron's energy adviser was a former British Gas lobbyist came from the *Independent* [37]. The story that at least eight energy industry employees have been seconded to DECC came from the *Guardian* [38].

The International Energy Agency World Energy Outlook provided the figures for government subsidies for fossil fuels and renewables world-wide [39]. It is not so easy to get this data for individual countries. The OECD provides data on fossil fuel subsidies in their member countries including the UK [40]. Strangely, the IEA only provides figures for renewable subsidies for the whole of Europe, not by individual countries [41]. I therefore used the REA values in reference 1 of Chapter 8 for the UK renewable subsidies.

It was a big step forward when the House of Commons Environmental Audit Committee commissioned evidence on subsidies from Dr William Blyth of Oxford Energy Associates. In contrast to the OECD–IEA situation, one can have more confidence that a single organisation, Oxford Energy Associates, will calculate renewable and fossil fuel subsidy figures with a consistent methodology.

Blyth's results (on page 26 of the report) show that the subsidy for gas (£3.6 billion) is bigger than for all the renewables together (£3.1 billion) [42]. I believe that this is important information for the environmental movement.

The International Energy Agency analysis of CO_2 emissions [43] shows 'electricity and heat generation' contribute 41 per cent and 'transport' contributes 22 per cent. Page 9 of this report emphasises the key point that 'the combined share of electricity and heat generation and transport represented nearly two-thirds of global emissions in 2010'.

1. Hardy Graupner, Deutsche Welle, 22 January 2013, http://www. dw.de/what-exactly-is-germanys-energiewende/a-16540762.

BIBLIOGRAPHY

2. Ms Stubholt, State Secretary, Norwegian Ministry of Petroleum and Energy, 11 March 2011, http://www.euractiv.com/energy/minister-norways-renewable-goals-interview-189823, accessed 18 January 2014.

3. EU Screening report Iceland, Chapter 15 – energy, 12 May 2011, http://ec.europa.eu/enlargement/pdf/iceland/key-documents/screening_report_15_is_internet_en.pdf, accessed 18 January 2014.

4. The Danish Government, 'Summary, Energy Strategy 2050: From Coal, Oil and Gas to Clean Energy', February 2011, http://marokko.um.dk/~/media/Marokko/Documents/Other/GBEnergistrategi2050sammenfatning.pdf, accessed 19 January 2014.

5. Preben Maegaard, 'Danish Renewable Energy Policy', 2009, http://www.wcre.de/index.php/presse/articles-mainmenu-17/119-danish-renewable-energy-policy, accessed 19 January 2014.

6. Fergus Ewing, Energy Minister, Scottish Parliament press release, 17 April 2012, http://www.scotland.gov.uk/News/Releases/2012/04/scottishrenewables17042012, accessed 5 February 2014.

7. Jonathan Leake, 'Green Energy Could Blot Out Countryside', *The Sunday Times*, 20 November 2011, http://thesundaytimes.co.uk/.

8. George Monbiot, *Heat: How to Stop the Planet Burning*, Penguin Books (2006).

9. George Monbiot, 'A Great Green Rip Off', *Guardian*, 1 March 2010, http://www.monbiot.com/2010/03/01/a-great-green-rip-off, accessed 19 January 2014.

10. UK House of Commons, Environmental Audit Committee, Memorandum submitted by Professor David MacKay, 9 July 2009, http://www.publications.parliament.uk/pa/cm200910/cmselect/cmenvaud/228/9071405.htm, accessed 19 January 2014.

11. Department of Energy and Climate Change (DECC), '2050 Pathways Analysis', July 2010, https://www.gov.uk/government/uploads/system/uploads/attachment_data/file/42562/216-2050-pathways-analysis-report.pdf, accessed 19 January 2014.

12. British Pugwash, 'Pathways to 2050: Three Possible UK Energy Strategies', 2013, http://www.britishpugwash.org/documents/British%20Pugwash%20Pathways%20to%202050%20INNERSREVsmall.pdf, accessed 19 January 2014.

BIBLIOGRAPHY

13. UK Government, Digest of UK energy statistics (DUKES), 5.10, 'Plant Loads Demand and Efficiency', https://www.gov.uk/government/publications/electricity-chapter-5-digest-of-united-kingdom-energy-statistics-dukes, accessed 19 January 2014.

14. US Department of Energy, 'FY 2014 Congressional Budget Request, Laboratory Tables', DOE/CF-0091, April 2013, http://energy.gov/sites/prod/files/2013/04/f0/Lab%20Table%20FY2014.pdf, accessed 19 January 2014.

15. UK National Renewable Energy Centre (Narec), 'Narec Solar Spun Out', 12 March 2012, http://www.narec.co.uk/media/news/Narec+Solar+Spun+Out_1008, accessed 19 January 2014.

16. Fraunhofer Institute for Solar Energy Systems (ISE), 'Data and Facts', 30 December 2012, http://www.ise.fraunhofer.de/en/about-us/data-and-facts, accessed 19 January 2014.
    Fraunhofer Institute for Wind Energy and Energy System Technology (IWES), http://www.iwes.fraunhofer.de/en.html, accessed 19 January 2014.

17. House of Lords, Select Committee on Science and Technology, 'Nuclear Research and Development Capabilities', 3rd report of Session 2010–2012, http://www.publications.parliament.uk/pa/ld201012/ldselect/ldsctech/221/22105.htm, accessed 6 February 2014.

18. World Record Academy, 'Most efficient single junction solar cell – QuantaSol sets world record', 4 July 2009, http://www.worldrecordacademy.com/technology/most_efficient_single_junction_solar_cell-Quantasol_sets_world_record_90269.htm, accessed 19 January 2014.

19. Bob May, 'Under-informed, over here', *Guardian*, 27 January 2005.

20. Helen Briggs, 'Science Faces Dangerous Times', BBC News, 30 November 2005, http://news.bbc.co.uk/1/hi/sci/tech/4482174.stm, accessed 5 February 2014.

21. Lord May, 'We Need More Nuclear Power Stations, Not Wishful Thinking', *Daily Telegraph*, 15 September 2004.

22. The Royal Society, 'The Royal Society's response to Department of Trade and Industry Review of UK Energy Policy', RS policy document 08/06, April 2006, http://royalsociety.org/uploadedFiles/

Royal_Society_Content/policy/publications/2006/8298.pdf, accessed 19 January 2014.

23. The Royal Society, 'Fuel cycle stewardship in a nuclear renaissance', October 2011, The Royal Society Science Policy Centre report 10/11, http://royalsociety.org/policy/projects/nuclear-non-proliferation/report/, accessed 19 January 2014.

24. Frank von Hippel, Rodney Ewing, Richard Garwin and Alison Macfarlane, 'Time to bury plutonium', *Nature*, **485**, 167 (2012).

25. Geoffrey Lean, 'Minister admits total failure of Sellafield MOX plant', *Independent*, 9 March 2008.

26. The Royal Society and the Royal Academy of Engineering, 'Shale gas extraction in the UK: A review of hydraulic fracturing', DES2597, June 2012, http://royalsociety.org/policy/projects/shale-gas-extraction/, accessed 19 January 2014.

27. International Energy Agency, 'World Energy Outlook 2012, Executive Summary' (2012), http://www.iea.org/publications/freepublications/publication/English.pdf, accessed 19 January 2014.

28. Robert Howarth, Renee Santoro and Anthony Ingraffea, 'Methane and the greenhouse-gas footprint of natural gas from Shale Formations', *Climatic Change*, **106**, 679, (2011).

29. Nathaniel Warner, Cidney Christie, Robert Jackson and Avner Vengosh, 'Impacts of Shale Gas Wastewater Disposal on Water Quality in Western Pennsylvania', *Environmental Science and Technology*, **47**, 11849 (2013).

30. House of Commons, *Hansard*, 16 December 2013, written answer, 'Nuclear Power Stations', column 406W, http://www.publications.parliament.uk/pa/cm201314/cmhansrd/cm131216/text/131216w0001.htm#13121634000015, accessed 21 January 2014.

31. Nuclear Decommissioning Authority, 'Annual Report and Accounts 2012–2013', 24 June 2013, http://www.nda.gov.uk/news/arac-2012-2013.cfm, accessed 19 January 2014.

32. David Toke's green energy blog is at: http://realfeed-intariffs.blogspot.co.uk/, accessed 5 February 2014.

33. First Solar press release, 'First Solar becomes first PV Company to produce 1GW in a Single Year', 15 December 2009, http://investor.firstsolar.com/releasedetail.cfm?ReleaseID=571605, accessed 19 January 2014.

BIBLIOGRAPHY

34. Terry Macalister, 'BP closes solar power business and blames global downturn', *Guardian*, 22 December 2011.

35. First Solar press release, 'First Solar Inc. announces fourth quarter and 2011 financial results', 28 February 2012, http://investor.firstsolar.com/releasedetail.cfm?releaseid=652462, accessed 19 January 2014.

36. BP, 'BP Financial and Operating Information 2007–2011', http://www.bp.com/content/dam/bp/pdf/investors/FOI_2007_2011_full_book.pdf.

37. Tom Bawden, 'No 10's New Energy Adviser Is a Former British Gas Lobbyist', *The Independent*, 23 May 2013.

38. Damian Carrington and Andrew Sparrow, 'Gas Industry Employee Seconded to Draft Energy Policy', *Guardian*, 11 November 2013.

39. International Energy Agency, 'World Energy Outlook 2012 Factsheet', http://www.worldenergyoutlook.org/media/weowebsite/2012/factsheets.pdf, accessed 5 February 2014.

40. OECD, 'OECD-IEA Fossil Fuel Subsidies and Other Support' (2013), (scroll down to 'Country Information' to download EXCEL file), http://www.oecd.org/site/tadffss/, accessed 5 February 2014.

41. Marco Baroni, Senior Energy Analyst, IEA, private communication.

42. House of Commons Environmental Audit Committee, Energy Subsidies, Ninth Report of Session 2013–14, Vol.1, 27 November 2013, http://www.publications.parliament.uk/pa/cm201314/cmselect/cmenvaud/61/61.pdf.

43. International Energy Agency, 'CO_2 Emissions from Fuel Combustion: Highlights' (2012), http://www.iea.org/co2highlights/co2highlights.pdf, accessed 20 January 2014.

11. Foretelling the Future

The quotation from Neils Bohr has recently been used by Allison Macfarlane, head of the US Nuclear Regulatory Commission [1]. In her case the quotation was, appropriately, applied to the long term safety of nuclear waste.

BIBLIOGRAPHY

David Mathisen's 1979 paper, which converted me to the need for a solar revolution, is well worth a read, given the accuracy of its predictions [2].

1. Matthew Wald, 'Head of U.S. Nuclear Watchdog emphasizes preparing for unknown', *New York Times*, 12 March 2013.
2. David A. Mathisen, '2079: A century of technical and socio-political evolution', *Impact of Science on Society*, **29**, 83 (1979).

12. The Electric Car

I took the history of the motor car in the US from the early part of Mary Bellis's account [1]. Amory Lovins' pioneering ideas about how fuel cell cars can be much lighter than a conventional car with a heavy internal combustion engine, while remaining as strong, can be found in his definitive *Scientific American* article. It can also be downloaded from Lovin's Rocky Mountain Institute website [2].

A useful handbook for potential purchasers of electric cars, describing the many variants, was published in 2010 by the Royal Automobile Club [3]. Developments are happening fast in this area, so some of the information may be out of date. David MacKay's book, which is reference 5 in Chapter 8, is a year older, but it contains some useful basic technical information on electric cars and their batteries.

Up-to-date information on electric cars in the UK can be found on websites like Next Green Car [4]. The site has information about the £5,000 UK government subsidy. Sadly, this does not make a great dent in the cost of the car. The website also gives up-to-date information on your nearest re-charging points.

Kevin Sharpe, founder of Zero Carbon World, the charity that is installing charging points in motels and restaurants, provided me with the encouraging information on his own experiences of charging an electric car by rooftop PV. The four miles for each kWh of electrical energy agrees with the figures in reference 3. The Zero Carbon World website is well worth a visit [5].

The figure that the sun shines for 844 hours a year in the UK on average was taken from figure 2b in reference 7 in Chapter 7. Strictly speaking, the figure is the average number of kWh of electrical energy generated by a PV system of 1 kW power facing south at the optimal roof angle.

The 844 hours is around 5 per cent less than Germany. Both in Germany and the UK the sun shines for less than 10 per cent of the number of hours in a year (8,760). This is why, when German PV was supplying 28 per cent of the power, it was only providing under 3 per cent of the energy (Chapter 7).

The average annual domestic mileage in the UK in 2010 was 8,430 miles per year [6].

My colleagues and I pointed out in reference 8 of Chapter 9 that QuantaSol technology will make possible both domestic use and additionally can power the family electric car for the average domestic mileage, even in the UK.

There is an account of the history and operation of the hydrogen fuel cell in the book written for a non-specialist audience by Peter Hoffmann [7]. Hyundai's plans to launch a hydrogen fuel cell in 2015 were taken from reference 8. Demonstration cars are already earmarked for EU parliamentarians and others have been ordered by Scandinavian cities.

1. Mary Bellis, 'The History of the Electric Car', About.com inventors, 27 August 2013, http://inventors.about.com/od/estartinventions/a/History-Of-Electric-Vehicles.htm, accessed 20 January 2014.
2. Amory Lovins, 'More Profit with less Carbon', *Scientific American*, **293**, 74 (September 2005), http://www.rmi.org/Knowledge-Center/Library/C05-05_MoreProfitLessCarbon, accessed 20 January 2014.
3. Arvid Linde, 'Electric Cars – The Future is Now!', published by Veloce Publishing Ltd for the Royal Automobile Club (2010).
4. Next Green Car, 'Electric Cars', http://www.nextgreencar.com/electric-cars/, accessed 20 January 2014.
5. Private communication, Kevin Sharpe, http://zerocarbonworld.org/.
6. UK Department for Transport, 'National Travel Survey 2010', 28 July 2011, http://www.dft.gov.uk/statistics/releases/national-travel-survey-2010, accessed 20 January 2014.

BIBLIOGRAPHY

7. Peter Hoffmann and Byron Dorgan, *Tomorrow's Energy: Hydrogen Fuel Cells and the Prospects for a Cleaner Planet*, Massachusetts Institute of Technology (2012).
8. Hyundai, 'Hyundai iX35 Fuel Cell Farm', http://www.hyundai.co.uk/ about-us/environment/hydrogen-fuel-cell, accessed 5 February 2014.

13. Solar Fuel

The methanol fuel cell electric car that is being developed in Denmark is described in an article on the Hybrid Cars website [1]. The 57 per cent efficient methanol fuel cell is already available from Serenergy [2] and I have had some extremely useful discussions with Mads Bang the CTO of that company [3].

A comparison of the merits of the 'methanol economy' and the 'hydrogen economy' for domestic and industrial processes and also transportation, plus their safety in comparison with petrol, can be found in the definitive book by the Nobel Prize-winner George Olah and colleagues [4]. For more information on the safety of methanol, see the US Methanol Institute website [5].

The American Physical Society (APS) report, which concludes that direct air capture of CO_2 with chemicals is not currently economically viable, can be downloaded from the APS website [6]. One of the projects in progress that the committee considered is described in a non-specialised news article as 'artificial trees' [7]. Reference 8 reviews technical issues concerning some approaches to direct carbon dioxide capture for sustainable fuels.

I have had many helpful discussions with Aldo Steinfeld [9]. His group is studying a number of approaches to extracting carbon dioxide and water from the air. One approach appears to be particularly suitable for powering with the waste heat of QuantaSol cells [10]. It cycles from ambient temperature up to 90 °C where we predict our cells will operate with 40 per cent efficiency [11].

Reference 12 contains details of a typical water-cooled CPV system. The system I propose would be a higher concentration and therefore a higher operating temperature. The difference in temperature between the cell and cooling water should be similar.

BIBLIOGRAPHY

I took Professor MacKay's values for the carbon dioxide generated by the average new UK car and the energy cost of carbon capture from pages 122 and 244 of his book which is reference 5 of Chapter 8.

I thoroughly recommend Jim Barber's description of photosynthesis [13], which I found fascinating even though I remember very little school chemistry and even less biology. My description of photosynthesis relies very heavily on Jim's and I am extremely grateful for the advice he has given me. The second figure in Jim's paper he describes as a 'simplified' schematic of photosynthesis. Jim credits Wikepedia for this figure, so it is easy to access. This 'Z scheme' figure brings home the staggering complexity of one of nature's most important achievements [14].

I have had a number of very helpful discussions about the use of biological cells to mimic photosynthesis with Bao-Lian Su [15] who I met at the same summer school as Aldo Steinfeld. A review of scientific papers in this area can be found in reference 16.

Art Nozik, whom I met at in my early visits to NREL, has co-edited an academic review of nanostructure approaches to the photoelectro-chemical cell (PEC) [17]. The artificial leaf is being studied in many laboratories worldwide, with much of the effort focusing initially on hydrogen generation. An old friend at Imperial, James Durrant [18], is one of those looking at solar fuel production from carbon dioxide. He recommended the review of this area by Michele Aresta and colleagues [19]. This in turn drew my attention to the work of Bocarsly and collaborators [20]. They claim to produce methanol directly from CO_2 using a pyridinium catalyst and have set up a spin-out company, Liquid Light to exploit this technology [21].

The international effort and expense in building the ITER fusion research reactor in the south of France is enormous [22]. Bear in mind that these funds for construction are in addition to the fusion research funds. In 2011–2012 the UK research council support for fusion research was £31.9 million and for PV research £8.8 million [23].

The US Department of Defense (DoD) were allocated $1.2 billion for biofuels, solar technology and batteries in 2011 [24] and a further $7 billion funding for procuring renewable energy [25].

1. Jeff Cobb, 'Range-extended QBEAK uses Bio-Methanol fuel cell', HybridCars, 9 July 2012, http://www.hybridcars.com/

pending-range-extended-qbeak-uses-bio-methanol-fuel-cell-48071/
accessed 5 February 2014.

2. Serenergy, The Power of Simplicity, Technology: HT PEM Basics,
19 August 2013, http://serenergy.com/technology/ht-pem-basics/.

3. Mads Bang, Chief Technical Officer, Serenergy, private
communication.

4. George Olah, Alain Goeppert and G.K. Surya Prakash, *Beyond Oil
and Gas: The Methanol Economy*, Wiley-VCH, 2nd edition (2009).

5. US Methanol Institute, 'Health and Safety' (2011), http://meth-
anol.org/Health-And-Safety.aspx, accesses 20 January 2014.

6. American Physical Society, 'Direct Air Capture of CO_2 with
Chemicals', 1 June 2011, http://www.aps.org/policy/reports/assess-
ments/upload/dac2011.pdf, accessed 20 January 2014.

7. Richard Schiffman, 'Artifical Trees as a Carbon Capture
Alternative to Geoengineering', The Yale Forum on climate
change and the media, 13 February 2013, http://www.yaleclimate-
mediaforum.org/2013/02/artificial-trees-as-a-carbon-capture-alter-
native-to-geoengineering/, accessed 5 February 2014.

8. Christopher Graves, Sune D. Ebbeson, Mogens Mogensen and
Klaus S. Lackner, 'Sustainable Hydrocarbon Fuels by Recycling
CO_2 and H_2O with Renewable or Nuclear Energy', *Renewable
and Sustainable Energy Reviews*, **15**, 1 (2011).

9. Aldo Steinfeld, Department of Mechanical and Process
Engineering, ETH Zurich, private communication.

10. Jan Andre Wurzbacher, Christoph Gebald, and Aldo Steinfeld,
'Separation of CO_2 from air by temperature-vacuum swing
absorption using diamine-functionalised silica gel', *Energy and
Environmental Science*, **4**, 3584 (2011).

11. K.W.J.Barnham et al., 'Photonic Coupling in Multi-Junction
Quantum Well Concentrator Cells', *Proceedings of the 5th
World Conference on Photovoltaic Energy Conversion*, WCPEC5,
Valencia (2010), p. 234.

12. X. Xu, M.M. Meyers, B.G. Sammakia and B.T. Murray, 'Thermal
Modeling and Life Prediction of Water-Cooled Hybrid concen-
trating Photovoltaic/Thermal Collectors', *Journal of Solar Energy
Engineering*, **135**, 011010 (2013).

13. James Barber, 'Photosynthetic energy conversion: natural and

BIBLIOGRAPHY

artificial', *Chemical Society Reviews*, **38**, 185 (2009).

14. Wikipedia, 'Z scheme' figure, http://en.wikipedia.org/wiki/Photosynthesis, accessed 5 February 2014.

15. Bao-Lian Su, Laboratory of Inorganic Materials Chemistry, The University of Namur (FUNDP), private communication.

16. A. Leonard, Ph. Dandoy, E. Danloy, G. Leroux, J.C. Rooke, C.F. Meunier and B.L. Su, *Chem. Soc. Rev.* **40**, 860 (2011).

17. Mary D. Archer and Arthur J. Nozik, eds., *Nanostructured and Photoelectrochemical Systems for Solar Photon Conversion*, Imperial College Press, (2008).

18. James Durrant, Department of Chemistry, Imperial College London, private communication.

19. Michele Aresta, Angela Dibenedetto and Antonella Angelini, 'The Use of Solar Energy Can Enhance the Conversion of Carbon Dioxide into Energy-Rich Products: Stepping Towards Artificial Photosynthesis', *Philosophical Transactions of the Royal Society A, Mathematical, Physical & Engineering Sciences*, **371**, 20120111 (2013).

20. Emily Barton, David Rampulla and Andrew Bocarsly, 'Selective Solar-Driven Reduction of CO_2 to Methanol Using a Catalyzed p-GAP Based Photoelectrochemical Cell', *Journal of the American Chemical Society*, **130**, 6342 (2008).

21. Liquid Light Corporation, 'Sustainable Chemicals from Carbon Dioxide', 2013, http://llchemical.com/, accessed 5 February 2014.

22. Steve Connor, 'One Giant Leap for Mankind: £13bn ITER project makes breakthrough in the quest for nuclear fusion, a solution to climate change and an age of clean, cheap energy', *Independent*, 27 April 2013.

23. House of Commons *Hansard*, 'Energy Expenditure', written answer, 4 February 2013, column 10W, http://www.publications. parliament.uk/pa/cm201213/cmhansrd/cm130204/text/130204w0001. htm#1302044000013, accessed 21st January 2014.

24. Felicity Carus, 'US Military Takes Solar to the Frontline', PV-TECH, 8 May 2012, http://www.pv-tech.org/editors_blog/us_ military_takes_solar_to_the_frontline, accessed 21 January 2014.

25. World of Renewables, 'US Army Receives $7B for Renewable Energy Projects', 14 August 2012,

BIBLIOGRAPHY

http://www.worldofrenewables.com/renewables_news/us_army_
receives_usd_7_billion_for_renewable_energy_projects.html,
accessed 21 January 2014.

14. Manifesto for the Solar Revolution

The Solar Manifesto is based on the arguments developed in *The
Burning Answer*. The relevant references can therefore be found earlier
in this bibliography. The only suggestion in this chapter that is not
discussed earlier is the idea of shareholder resolutions at the larger
fossil fuel companies. This approach was effective in restricting loans
to the South African government at the time of apartheid. More details
of the successful shareholder actions at that time can be found in the
account written by one of my oldest friends, Reverend David Haslam
[1]. David initiated me into a lifetime of support for causes unpopular
with the rich and powerful.

An internet campaigning group that has taken on a major bank is 350.
org, which is campaigning to get the European Bank for Reconstruction
and Development to stop funding new coal plants [2].

Operation Noah is an active group, challenging the churches over
their investments in fossil fuel industries [3].

The World Development Movement is taking on the banks, pension
funds and finance companies that are funding dirty energy projects [4].

The Smith School of Enterprise and the Environment has recently
produced a report [5] with useful information on university sharehold-
ings in fossil fuel companies. I believe that their conclusion 'the direct
impacts of fossil fuel divestment on equity or debt are likely to be
limited' is missing the point. In the case of apartheid, a relatively small
shareholder revolt resulted in some banks stopping investing in the South
African government and its state enterprises. Campaigns should try to
persuade fossil fuel companies not to develop unpopular, unconventional
sources and instead focus on geothermal energy, biogas from waste, or
research and development on low carbon fuels. Such campaigns might
be successful, even though resolutions are only supported by a minority
of shareholders.

A particularly useful resource for such a campaign is the report

BIBLIOGRAPHY

written by Richard Heede for the Climate Accountability Institute. His calculations enable him to identify the top 20 greenhouse-gas emitting companies and state enterprises. He also calculates that 63 per cent of the cumulative greenhouse gas emissions in the industrialised world have been from just 90 companies or state enterprises. The report can be downloaded from the Climate Accountability Institute [6].

1. David Haslam, 'Mobilizing the European Churches', Chapter 5 in *A Long Struggle: the Involvement of the World Council of Churches in South Africa*, ed. Pauline Webb, WCC publications, Geneva (1994).
2. 350.org, 21 August 2013, http://act.350.org/sign/EBRD/.
3. Operation Noah, 'Faith motivated, science informed, hope inspired', 10 December 2013, http://www.operationnoah.org/, accessed 21 January 2014.
4. World Development Movement, 'Carbon Capital', http://www.wdm.org.uk/carbon-capital, accessed 21 February 2014.
5. Smith School of Enterprise and the Environment, University of Oxford, 'Stranded assets and the fossil fuel divestment campaign: What does divestment mean for the valuation of fossil fuel assets', June 2013, http://www.smithschool.ox.ac.uk/research/stranded-assets/SAP-divestment-report-final.pdf.
6. Richard Heede, 'Tracing Anthropogenic Carbon Dioxide and Methane Emissions to Fossil Fuel and Cement Producers, 1854–2010', Climatic Change 122, 229–241 (2014). Report can be downloaded from: http://www.climateaccountability.org/publications.html, accessed 22 January 2014.

15. Tomorrow the World

My description of the parts of the world that have twice the amount of sunlight energy in a year compared to Britain was based on the top figure on the Green Rhino Energy site [1]. On the world map of the yearly sums of global irradiance, the south-west UK is red and the rest of the country is purple. The key suggests a UK average around 1000 kWh per square metre in a year. Large areas of land closer to the

BIBLIOGRAPHY

equator than the UK are yellow and so have more than 2000 kWh per square metre in a year.

Information about the DESERTEC project can be found on their website [2].

Massimo Mazzer, Barry Clive and I reported the rise in air conditioning demand in the EU in an article in *Nature Materials* in 2006 [3]. Air conditioning demand has continued to rise in the EU despite a hiatus around the time of the 2008 credit crunch [4]. A typical electricity demand for California shows no daytime peak but an overall rise to a peak around 6.00 p.m. [5]. The afternoon demand in Germany and the UK is much lower, particularly in summer, as discussed in Chapter 7. This suggests air conditioning is a major component of demand in places like California. Summer electricity demand in Italy also shows evidence for an air conditioning component as discussed in reference 4 of Chapter 7. I also like reference 5 because the paper shows the results of Amonix field tests around the south-west USA using triple junction cells from a number of suppliers that included QuantaSol. Of the cells tested the QuantaSol cell turned out to be best optimised for the Amonix concentrator, thanks to the quantum wells.

Our *Nature Materials* article also suggested solar-powered factories in 2006 [3].

The new Italian company Film4Sun developing window shutters that generate electricity for air conditioning can be visited at reference 6.

I think it important for the further development of third-generation CPV in general, not just in its application to high latitudes, that one of the first high concentration CPV systems deployed in a coastal region of Japan survived a typhoon [7].

Nicoletta Marigo had the bright idea for her PhD project at Imperial; a study of the potential of PV in China. When she started, there was hardly any Chinese PV industry. By the time she had completed her thesis, most of the world-leading PV companies were Chinese [8]. She has provided me with much helpful information on PV and other energy developments in China.

Nicoletta emphasises that the expansion of PV manufacturing in China was not only due to strong government support, but also to ensuring strong material supply chains. The resulting dramatic fall in the price of first and second generation panels, discussed in Chapter 8, was bad news for Western PV manufacturers, including third-generation companies like QuantaSol. Another consequence was the introduction of protectionist measures in the US and EU [9], which I fear could slow the pace of PV expansion.

More positively, the Chinese companies are now selling strongly in their own, domestic market and recently announced a massive $40 billion investment in PV [10].

A number of companies are looking at deploying small third-generation CPV cells the size of LEDs in concentrator systems. Many of the manufacturing processes are compatible with LED mass production techniques. One approach is being taken by my old colleague from Philips, Geoff Duggan [11]. Carlos Algora of the Polytechnic University of Madrid has a different approach [12].

The official history of CERN from inception immediately post-war to the discovery of the Higgs boson can be found on the CERN website [13].

Rotblat's idea about the need to modify the bargain in the Non-Proliferation Treaty that encourages the nuclear haves to co-operate with the nuclear have-nots is in reference 14. His suggestion of a world solar laboratory was also made in that chapter.

The IAEA budget and the 38 per cent proportion for safeguards work can be found in reference 15. If one adds on the equally important safety and security budget more than half the remainder of the budget is for research, development and the promotion of nuclear power. A world solar laboratory should aim for similar support from the international community.

The International Renewable Energy Association (IRENA) is much newer and very much smaller than the IAEA. As far as one can tell from the IRENA website [16], it has about 21 employees. The IAEA has a staff of 2,300 [17]. As the only worldwide body promoting all the new and important solar industries I feel IRENA deserves larger support from the international community for its promotional and collaborative work.

Another positive aspect of the IAEA budget is the support it gives to the International Centre for Theoretical Physics (ICTP) founded by the late Abdus Salam, which supports physicists from the developing world [18].

BIBLIOGRAPHY

My ideas about potential benefits should solar become the 'new nuclear' are expanded further in a paper I wrote for Scientists for Global Responsibility in July 2012 [19]. The paper discusses the Iran situation in detail and references possible Saudi Arabian interest in the nuclear option. My concern about the links between civil and military nuclear activities on the world scale, and the possibility of their mitigation by renewable energy, were first prompted by reading Joseph Rotblat's chapter in reference 14, and also Amory and Hunter Lovins' book [20]. These were published in 1979 and 1980 respectively; around the time David Mathisen made his perceptive forecasts that I referred to in Chapter 11.

1. Green Rhino Energy, 'Yearly Sum of Global Irradiance' http://www.greenrhinoenergy.com/solar/radiation/empiricalevidence.php, accessed 5 February 2014.

2. Desertec Foundation, 8 January 2014, http://www.desertec.org/ accessed 5 February 2014.

3. K.W.J. Barnham, M. Mazzer and B. Clive, 'Resolving the Energy Crisis: Nuclear or Photovoltaics?' *Nature Materials*, 5, 161 (March 2006).

4. Paolo Bertoldi, Bettina Hirl, Nicola Labanca, 'Energy Efficiency Status Report 2012', European Community Joint Research Centre, Ispra, Italy, JRC69638, 2012, http://iet.jrc.ec.europa.eu/energyefficiency/sites/energyefficiency/files/energy-efficiency-status-report-2012.pdf.

5. Geoffrey S. Kinsey, Aditya Nayak, Mingguo Liu and Vahan Garboushian, 'Increasing Power and Energy in Amonix CPV Solar Power Plants', *IEEE Journal of Photovoltaics*, 1, 213 (2011).

6. Film4Sun, 'Advanced Solution for Photovoltaics', http://www.Film4sun.com/, accessed 21 January 2014.

7. Kenji Araki, Daido Steel Company, private communication.

8. Nicoletta Marigo, 'Innovating for Renewable Energy in Developing Countries: Evidence from the Photovoltaic Industry in China', PhD thesis, Centre for Environmental Policy, University of London – Imperial College London (2009).

9. John Parnell, 'EU-China Dispute to Drive European Solar Firms to Bankruptcy: IHS', PV- Tech, 24 July 2013, http://www.pv-tech.

org/news/29024, accessed 21 January 2014.

10. Julia Chan, 'China to invest US$39.5B in PV during 2011–15', PV-Tech, 17 September 2012, http://www.pv-tech.org/news/21535, accessed 21 January 2014.

11. Geoff Duggan, Full Sun Photovoltaics, private communication.

12. Carlos Algora, 'High Efficiency Photovoltaic Converter for High Light Intensities Manufactured with Optoelectronic Technology', European and US patent applications EP1278248A1 and US2002/0170592A1 (2002).

13. CERN, 'The History of CERN', http://timeline.web.cern.ch/time-lines/The-history-of-CERN, accessed 21 January 2014.

14. Joseph Rotblat, Chapter 14 in *Nuclear Energy and Nuclear Weapon Production*, SIPRI, Taylor and Francis (1979).

15. International Atomic Energy Agency (IAEA), 'IAEA Regular Budget for 2014', http://www.iaea.org/About/budget.html, accessed 21 January 2014.

16. International Renewable Energy Agency (IRENA), http://www.irena.org/home/index.aspx?PriMenuID=12&mnu=Pri, accessed 21 January 2014.

17. International Atomic Energy Agency 'Employees and Staff: Strength through Diversity', http://www.iaea.org/About/staff.html, accessed 21 January 2014.

18. International Centre for Theoretical Physics (ICTP), http://www.ictp.it/, accessed 5 February 2014.

19. Keith Barnham, 'Breaking the Deadlock: Iran's Nuclear Programme in Context', *SGR Newsletter*, Autumn 2012, Issue 41, http://www.sgr.org.uk/publications/sgr-newsletter-41, accessed 21 January 2014.

20. Amory B. Lovins and L. Hunter Lovins, *Energy/War: Breaking the Nuclear Link*, Harper Colophon Books (1980).

16. Wishful Thinking?

The YouGov survey, from which the quotation at the head of this chapter was taken, was requested by Friends of the Earth [1]. It had a sample size of 2,884 British adults. When asked which single energy source they would like to see in ten years' time only 3 per cent said 'coal' and only 2 per cent answered 'gas'.

By May 2013 the UK had installed 3.3 GW of offshore wind power. The average rate of increase in electrical energy generation was 54 per cent over the previous eight years [2]. The workers and companies concerned deserve great credit. This is similar to the average increase in PV power in Germany 2005 to 2010 in much easier working conditions.

The decision to halt the development of the Atlantic Array offshore wind farm was reported in the *Guardian* [3] where a cost of £4 billion was quoted. The potential power generation from the Atlantic Array site 1.5 GW was quoted in the 2012 version of reference 2. The cost of one of the two 1.6 GW reactors intended for Hinkley Point was estimated at £8 billion [4]. The excellent performance of Denmark's offshore wind turbines in recent years, approaching that of the current UK nuclear reactors (reference 13 in Chapter 10), is reported in reference 5.

The Wymeswold solar farm in Leicestershire has 33 MW of PV over a 150-acre site [6]. The Hinkley Point C reactor size of 438 acres was taken from reference 7.

Very sensibly, while waiting for the prototype EPR to be completed before starting to build the EPR at Hinkley Point, Areva has been diversifying into offshore wind. Since 2004 the French company has been using its expertise in the manufacture of large steam turbines for nuclear reactors to deliver wind turbines to the North Sea [8].

The recent measurements on the shrinking of the Arctic icecap are described by Ramez Naam on the *Scientific American* website as a 'triple whammy' [9]. They confirm global warming, there is less ice reflecting back sunlight further upsetting the delicate balance described in Chapter 7 and the unfrozen and exposed ground is now leaking more methane into the atmosphere.

The average carbon dioxide level reached the level of 400 parts per million in late May 2013, up from 320 parts per million in 1960 [11]. There is a report of an interview with Ralph Keeling, director of the Scripps CO_2 Program, on the significance of this level available on the *Guardian Environment* website [10]. Ralph is the son of Charles Keeling, who found an accurate way to measure this relatively small proportion of CO_2 in the atmosphere in 1958. You can follow day-by-day measurement of CO_2 level in the atmosphere from Mauna Loa Observatory on the Keeling Curve site at Scripps Institution of Oceanography [11].

The 2013 report from the Intergovernmental Panel on Climate Change (IPCC), which presented the convincing evidence that humankind is responsible for most of the global warming, is reference 12. If you, or anyone you are discussing with, still doubts this, compare two graphs: Figure SPM.1a in the IPCC report (the measured rise in global temperature averaged over a decade) and the Keeling curve graph for years 1700 to the present date. Both of them start an upward trend around 1900 when oil started to be seriously exploited. The carbon dioxide graph starts rising faster around the 1960s. The temperature increase follows the carbon dioxide level with the steeper rise starting from the decade 1971 to 1980.

1. YouGov, 'Survey Results', Fieldwork 16–17 April 2012, http://d25d2506sfb94s.cloudfront.net/cumulus_uploads/document/rq2xuw3tr8/YG-Archives-FriendsofTheEarth-Electricity-230412.pdf, accessed 21 January 2014.
2. Crown Estate, 'Offshore Wind Operational Report 2013', http://www.thecrownestate.co.uk/media/418869/offshore-wind-operational-report-2013.pdf, accessed 21 January 2014.
3. Terry Macalister, 'Energy Firm RWE npower Axes £4bn UK Windfarm Amid Political Uncertainty', *Guardian*, 26 November 2013, http://www.theguardian.com/environment/2013/nov/26/renewable-energy-rwe-drops-uk-turbine-project, accessed 22 January 2014.
4. Sean Farrell, 'Hinkley Point: nuclear power plant gamble worries economic analysts', *Guardian Environment*, 30th October 2013, http://www.theguardian.com/environment/2013/oct/30/hinkley-point-nuclear-power-plant-uk-government-edf-underwrite, accessed 5 February 2024.
5. Energy Numbers, 'Capacity Factors at Danish Offshore Wind Farms', 21 January 2014, http://energynumbers.info/capacity-factors-at-danish-offshore-wind-farms, accessed 22 January 2014.
6. Damian Carrington, 'Solar Farm Owners Reject Tory Monster Claims', *Guardian*, 23 December 2013.
7. EDF Energy, 'Hinkley Point C: an Opportunity to Power the Future', February 2013, http://www.edfenergy.com/about-us/energy-generation/new-nuclear/hinkley-point-c/book/book/index.

html#/4/, accessed 22 January 2014.

8. Renewable Energy Focus, 21 January 2014, http://www.renewableenergyfocus.com/view/36555/gamesa-and-areva-join-forces/, accessed 5 February 2014.

9. Ramez Naam, 'Arctic Sea Ice: What, Why and What Next?' *Scientific American*, 21 September 2012, http://blogs.scientifi-camerican.com/guest-blog/2012/09/21/arctic-sea-ice-what-why-and-what-next/, accessed 22 January 2014.

10. Fen Montaigne, 'Record 400ppm CO_2 Milestone "feels like we're moving into another era"', *Guardian Environment*, 14 May 2013, http://www.theguardian.com/environment/2013/may/14/record-400ppm-co2-carbon-emissions, accessed 22 January 2014.

11. Scripps Institution of Oceanography at University of California, 'Keeling Curve', http://keelingcurve.ucsd.edu/, accessed 21 January 2014.

12. IPCC, 'Climate Change 2013: The Physical Science Basis', Summary for Policymakers, October 2013, http://www.climat-echange2013.org, accessed 21 January 2014.

Index

INDEX

INDEX

INDEX

INDEX

INDEX

INDEX

INDEX